COMMUNICATION

PHYSICAL SCIENCE AND TECHNOLOGY SERIES

Editors

Karl K. Turekian *Yale University*
J. Rimas Vaišnys *Yale University*

Avizienis	INFORMATION AND COMPUTERS
Bernstein	LAWS OF NATURE
Cameron	STRUCTURE OF THE UNIVERSE
Davidovits	COMMUNICATION
Davidovits	TRANSPORTATION
Fenn	TRANSFORMATIONS OF ENERGY
Gordon	PHYSICS OF THE EARTH
Turekian	CHEMISTRY OF THE EARTH
Vaišnys	CHEMISTRY AND PHYSICS OF LIFE
Wheeler	STRUCTURE OF MATTER
Willis	RELATIVITY AND UNCERTAINTY

COMMUNICATION

PAUL DAVIDOVITS *Yale University*

HOLT, RINEHART AND WINSTON, INC.

New York Chicago San Francisco Atlanta
Dallas Montreal Toronto London Sydney

Copyright © 1972 by Holt, Rinehart and Winston, Inc.
All rights reserved
ISBN: 0-03-084784-2
Library of Congress Catalog Card Number: 70-170943
Printed in the United States of America
2 3 4 5   0 9 0   9 8 7 6 5 4 3 2 1

Editors' Foreword

hysical science and technology in themselves bring neither salvation nor doom r mankind. Rather, they are as much a part of man's cultural environment as oetry and music. The comprehension of the physical world and its harnessing re no less acts of the human spirit than the creation or appreciation of works f art.

This series presents fundamental knowledge of our physical world in as elementary a way as the editors and each author thought possible, without reducing ne presentation to banality. Not all parts of the books can be read with equal ease. Iowever, just as there is no effortless way to learn to play the violin well, there no way to comprehend nature without applying oneself to the task. It is hoped nat through these books the reader will not only discover the beauty of the hysical universe, but will also learn about the processes that lead to an undertanding of the universe.

The series format permits teachers and students to choose the material to suit neir own aims and interests.

KARL K. TUREKIAN
J. RIMAS VAIŠNYS

Editors

Preface

ll aspects of our society have been profoundly influenced by the applications
science. It is now evident that the results have been mixed. In many ways
ience has improved the quality of our lives. In scientifically-oriented societies
e standard of living has risen. People live longer, have more leisure time, and
ve a greater opportunity to develop their potential. On the other hand, with
ience, man's destructive power has increased to the point where he can an-
hilate himself. We now live in a world which is continuously threatened, if not
a cataclysmic destruction, then by a slow pollution and erosion of our environ-
ent. We are surrounded by gadgets and techniques that most of us do not
derstand.

We are beginning to realize that technology left to follow its own random
urse creates more problems than it solves. It is becoming evident that many
ientific and technological feats which at first appear most impressive are in
e long run undesirable and destructive. To avoid chaos, we must become aware
the potential and the dangers of science, and we must participate in guiding
course. But this has to be done intelligently with knowledge and understanding.

How can one become scientifically and technologically aware and literate?
he obvious answer is to study science and technology. I do not think that it is
cessary for all to acquire the specialized expertise of a professional scientist.
ut I do think that an understanding of the methods, techniques, and potentials
science and technology is essential to the effectiveness of an individual in
r society.

The usual science education is confined to a study of basic physical principles
d seldom includes an examination of the applications of science to technology.
think such an education is now inadequate. Certainly I think basic science
ould be taught, but somewhere along the line the applications and implications
science should be examined in greater detail.

This book describes the evolution and operation of communication technology.
s aim is to acquaint the reader with the development and functioning of a

technology which has been one of the most influential in shaping society. previous scientific background is not necessary for the understanding of t book. All basic scientific concepts underlying the technology are explained the text, in most cases without the use of mathematical formalism.

The first chapter describes the methods of communications used before t invention of electrical communication systems. In the second chapter ele tricity, magnetism, and their interactions are explained. This chapter also co tains a description of the Bohr atom and explanations of the operation of son devices which resulted from the early discoveries and were subsequently us in communication systems. Chapters 3 through 9 describe communication sy tems familiar to most readers. In these chapters the basic principles of develo ment and operation are explained for telephone, telegraph, radio, sound repr duction and television. Here, as in the rest of the book, the discussion is simp and self-contained. Wherever possible the text maintains an historical perspecti and includes the biography of some of the leading workers in the field.

The remaining part of the book deals with less well-known aspects of co munication. The operation of semiconductor devices (diodes, transistors, a so forth) is explained in Chapter 10. High frequency communication techniqu including multiplexing and satellite communication are described in Chapter 1 In Chapter 12 the theory and operation of lasers are explained. Lasers ha important potential applications in communications. Some of these applicatio are discussed.

In Chapter 13 the trends in communication technology are discussed and son of the forthcoming devices and developments are described.

The effects of communication technology on society are not discussed this book. However, a bibliography on various nontechnological aspects communications is included in Appendix I.

Appendix II is an evaluation of extrasensory perception (ESP). This subje is controversial. Some people believe that ESP is a real phenomenon; others not. Since many proponents of ESP have suggested that it could be used as practical communication means, I decided that its evaluation in this book appropriate.

Review problems for the text are given in Appendix III.

P.D.
New Haven, Con
November 1971

Contents

COMMUNICATION

1 Pretechnological Communication

The methods of communication from prehistory until the invention of the electric telegraph changed very little. People communicated by visual means, by sound, and most frequently by messengers. Aeschylus in his play *Agamemnon* records that the fall of Troy was signaled to Greece by prearranged fires on tops of mountains between Greece and Troy. The watchman stationed on the roof of King Agamemnon's palace in Argos speaks:

> I wait; to read the meaning in that beacon light,
> a blaze of fire to carry out of Troy the rumor
> and outcry of its capture;

And when the light is seen he says:

> I cry the news aloud to Agamemnon's queen,
> that she may rise up from bed of state with speed
> to raise the rumor of gladness welcoming this beacon
> and singing rise, if truly the citadel of Ilion
> has fallen, as the shining of this flare proclaims.

Aeschylus, Agamemnon[1]

Agamemnon's signaling system was capable of relaying one prearranged message. As such it was quick and simple but obviously very limited.

One of the earliest signaling systems capable of transmitting full alphabetical information was designed in 300 B.C. by Polybius, a Greek general. Polybius built transmitting stations consisting of two walls about 3 feet apart. The walls were about 7 feet long and 6 feet wide. Letters of the alphabetical message were transmitted by prearranged combinations of torches on the wall. The code is shown in Table 1-1. For example, two torches on the right wall and three torches on the left wall signaled "Theta" (Figure 1-1). This method of signaling differed very little from the semaphore systems which were used extensively through the nineteenth century and which, to some extent, are still used today.

---

[1] *Greek Plays in Modern Translation,* edited by Dudley Fitts. New York: Dial Press, 1947. With permission by Richmond Lattimore, translator.

Table 1-1   Code Used by Polybius, 300 B.C.

		Left			
	1	2	3	4	5
	1	A	B	Γ	Δ	E
	2	Z	H	Θ	I	K
Right	3	Λ	M	N	Ξ	O
	4	Π	P	Σ	T	Y
	5	Φ	X	Ψ	Ω	

SOURCE: *Data Transmission*, W. R. Bennett and J. R. Davey. New York: McGraw-Hill Book Company, 1965. Used with permission of McGraw-Hill Book Company.

Sound communication is, of course, also ancient. Since prehistoric times trumpets, gongs, and drums have been used to warn of danger and to call assemblies. Signaling with drums has been highly developed in parts of Africa. Different drums are associated with particular messages. For example, the Akan people of Ghana use a very high-pitched drum to call the people for a meeting at the chief's house. The African "talking" drums are especially interesting. These are actually constructed to imitate human speech by transmitting its tonal pattern. Since many words have the same tonal pattern, long redundant sentences have to be transmitted in order to make the message unambiguous. The transmission of a single message may last as long as 30 minutes. For example, the message, "The missionary coming upriver to our villa tomorrow, bring water and firewood to his house," transmitted in the drum language of the Kalai as

> White man spirit from the forest of the leaf used for roofs comes upriver, come upriver, when tomorrow has risen on high in the sky to the town and the village of us. Come, come, come bring water of [a specific type] vine, bring sticks of firewood to the house with shingles high up above of the white man spirit from the forest of the leaf used for roofs.[2]

The sound of these drums can be heard as far as 10 miles from the source. Important messages are often relayed from village to village, but intercultural drum communication among peoples with different languages is rare.

The American Plains Indians also had a highly developed signaling system. Colonel Dodge in his book, *The Hunting Grounds of the Great West*, published in 1877, wrote:

> In communicating at long distances on the plains, their mode of telegraphing is . . . remarkable. Indian scouts are frequently employed by the United States Government, and are invaluable, indeed almost indispensable, to the success of important expeditions. The leader, or interpreter, is kept with the commander of the expedition, while the scouts disappear far in advance or on the flanks. Occasionally one shows himself, sometimes a mere speck on a distant ridge, and the interpreter will say at once what that scout wishes to communicate . . .[3]

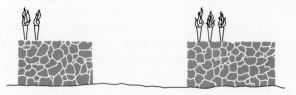

Figure 1-1   Signaling system used by Polybius, 300 B.C.

[2]*Communication in Africa*, Leonard W. Doob. New Haven, Conn.: Yale University Press, 1961.

[3]*The Great Plains*, Walter Prescott Webb. New York: Grosset and Dunlop, 1931.

Throughout history messenger pigeons have been in common use. This breed of pigeon has the ability to return home from any point of release. They have returned from distances as far as 2000 miles. The following is a filler paragraph from the *New Haven Register*:

A homing pigeon owned by the first duke of Wellington dropped dead a mile from its loft at Nine Elms in London, on June 1, 1845, 55 days after it was released from a sailing ship off the Ichabo Islands, West Africa, 5400 miles away.

There are reports of pigeons having been used as messengers in the days of King Solomon. The ancient Persians, Greeks, and Romans used them to deliver military messages.

In World War I pigeons were used extensively by both sides. During the Battle of the Somme, the French alone used about 5000 pigeons. On occasion, pigeons were used even during World War II. The British obtained information from their agents behind enemy lines by parachuting pigeons in small containers to the agents: the pigeons were then released with coded messages tied to their feet.

Still, most long-distance communication was done by human messengers. In most cases the messenger services were slow and not very reliable but a few of them were remarkably efficient. During the height of the Roman power over 50,000 miles of road connected the Empire. There were rest stations and inns every 20 to 30 miles along the road where a traveler could eat, rest, stay overnight, and obtain fresh horses. Travel along these roads was safe and relatively fast. A message could be delivered from Rome to London in 13 days. This distance could not be covered any faster until the introduction of mechanized transportation.

One of the most highly developed pretechnological messenger services was the Pony Express, started on April 3, 1860, by a group of investors (Figure 1-2). Messengers in relays delivered mail from St. Joseph, Missouri, to Sacramento, California, a distance of nearly 2000 miles. During the trip, which took about 10 days, 75 horses were used. Trips were made once a week and the initial price for delivery was $5.00 per half ounce. The price was gradually reduced to about $1.00 per half ounce. Each man rode between 75 and 100 miles before he was relieved. Between 50 and 80 riders were employed at a salary of $50.00 a month.

Figure 1-2   The pony express. (From a painting by George M. Ottinger, courtesy of the Library of Congress.)

The trail passed through territory inhabited by hostile Indians. Although ther were many instances of riders chased by Indian war parties, the riders were usuall able to escape because their horses were better fed and faster than those of th Indians. The biggest danger from Indians was not to the riders but to the keepers o pony stations along the route. These stations were often attacked, the keepers killed and the horses stolen. But it was technology and not the Indian attacks that cause the collapse of the Pony Express. With the help of the United States Governmen a telegraph line was being built between the East and the West. When the line wa completed, the Pony Express became obsolete. On October 25, 1861, a little ove a year after it started, the Pony Express was discontinued.

By the late 1700s state and military affairs were so complex that the traditiona communication with messengers became inadequate. The optical semaphore systems which were developed about this time solved the problem at least tempo rarily. The most widely used semaphore system was the one designed by Claude Chappe (1763-1805). Chappe constructed the first model of the semaphore while he was a student at a seminary in Angers, France. His two brothers were studyin, at another seminary about a mile away and he built the system to communicate witl them. The device consisted of a wooden beam with movable arms at each end The relative position of the arms represented letters of the alphabet, numbers, an phrases. Chappe perfected his invention and after much struggle he convinced the French Government to try it. This was during the year 1793. The republican revo lution was at its height and the French forces were assailed by the allied armies o Britain, Holland, Prussia, Austria, and Spain. France had one advantage: the at tacking forces were badly coordinated. At this point it was vital for France to have a fast, reliable communication system. The government adopted the Chappe system and in 1794 built a 144-mile communication link between Paris and Lille. Stone towers separated by distances of 6 to 10 miles were built between the two cities The semaphore device was placed on top of the tower. A drawing of the tower and some of the symbols used are shown in Figure 1-3. The cross arm was 14 feet long and the pivoted arms were 6 feet. By means of ropes and pulleys the orientation of the bars could be varied to produce 196 distinct signals. Telescopes were used for signal observation. The first telegram was transmitted from Lille to Paris or August 5, 1794. It informed the government that the French forces had retaken Le Quesnoy. The Chappe telegraph system appears to have had a major role in revers ing the course of the war and turning it in favor of the French.

Soon many other telegraph systems were built throughout Europe. In Denmark a line was built in 1802. Another link was constructed in France between Paris and Toulon. This 475-mile system had 120 stations. The transmission of a signal from Paris to Toulon took between 10 and 12 minutes, but the rate was slow—about one signal a minute. In England stations were constructed linking the south coast with London to warn of a French invasion. The most ambitious project was built in Russia under Czar Nicholas I. Here 220 stations connected the 1200 miles from the Prussian frontier via Warsaw to St. Petersburg. The system required 1300 operators.

In Europe these communication systems served only the government. They were used mostly to relay military information and orders. On the other hand, in the United States semaphores connected main ports with inland cities to notify subscribers of ship arrivals. In 1804 for a fee of about $18 a subscriber in Boston would be notified that a ship had arrived from Europe. Although the fee was high it was probably worth it to be able to sell the old stock before the new merchandise arrived.

Figure 1-3 Chappe's semaphore in action during the siege of Conde-sur-l'Escaut, November 1794 with some typical signals shown below. An early example of mechanical signaling for tactical purposes in battle. (Courtesy of the Bettmann Archive.)

The semaphore systems had many inherent faults. Communication was not possible during fog, the signals could be read and decoded by the enemy, and, above all, the transmission of messages was slow. Because of the large number of stations and operators required, the system was very expensive to maintain. In 1837 the United States Congress was considering the construction of a semaphore between New York and New Orleans. They asked for advice from military men and scientists. They received an opinion from S. F. B. Morse that electrical telegraphing would soon make optical semaphores obsolete. The prediction was correct. By 1850 most long-distance telegraphy was done with electricity.

2 Electricity, Magnetism, and Their Interaction

Modern communication is based almost entirely on the application of electromagnetic phenomena. A study of the technological aspects of communications must therefore begin with an explanation of these effects. This is a full subject in itself and has been treated rigorously in many textbooks (see References). Here our description will be brief and only qualitative; however, it should be adequate for the understanding of subsequent topics.

INTRODUCTION TO ELECTRICITY

The first reports of electric phenomena come from ancient Greece. It was known then that when amber is rubbed against a cat's fur, the amber attracts small objects before touching them. This property of amber disappears a few minutes after the rubbing. The phenomenon was thought to be unique to amber and was named *electricity* after the Greek word for amber which is "electros." A more substantial knowledge of electricity was not obtained until about the year 1600 when William Gilbert (1540-1603) showed that electricity is not a unique property of amber but that it is manifest in many other substances. By the middle of the 1700s, through the work of Gilbert, Otto van Guericke (1602-1686), DuFay (1698-1739), Pieter van Musschenbroeck (1692-1761), Stephen Gray (1696-1736), and many others a large amount of phenomenological knowledge was obtained about electricity. The extent of the understanding can best be shown through a few simple experiments (Figure 2-1). We shall describe these experiments in contemporary terminology, although the full meaning of the terms will not be explained until later.

All electric phenomena are the result of interactions of electric charges. Later we shall describe the nature of these charges, but at this point we only note that they can be placed on objects and can be transferred from one object to another.

In our experiment a small dry object, such as a light ball made of pith, is suspended from a silk thread and an amber rod is electrified by rubbing it with cat's fur. When the amber rod is brought near the pith ball, the ball is attracted to the rod but after touching the rod, the ball is repelled by it. Through contact with the elec

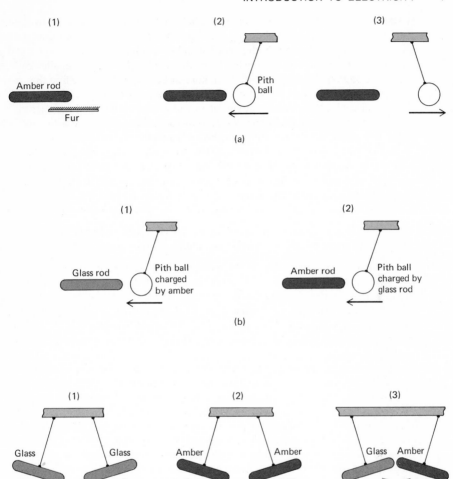

Figure 2-1 Early experiments with electricity. (a) An amber rod which has been rubbed by cat's fur (1) first attracts a pith ball (2), but after the ball touches the rod, the ball is repelled by the rod (3). The same result can be produced by a glass rod which has been rubbed by silk. (b) A pith ball which has been charged by amber is attracted by the glass rod (1), and a pith ball that has been charged by glass is attracted by the amber rod (2). (c) Two charged glass rods and two charged amber rods repel (1, 2), but a charged glass rod and a charged amber rod attract (3).

trified amber, the pith ball has also become electrified. In today's terminology we would say that in the course of rubbing, the amber has acquired electric charges and has become charged. The uncharged pith ball is initially attracted by the charged amber. When the pith ball touches the amber, some of the charges from the amber are transferred to the pith ball, causing it to become charged. The pith ball and the amber now have the same type of charge and they repel each other.

The same behavior can be observed by touching the pith ball with a glass rod rubbed with silk. But there is a difference between the pith ball charged by the amber rod and the one charged by the glass rod. The pith ball which has been charged by the amber rod is then repelled by the charged amber, but it is attracted to a charged glass rod. Similarly, the pith ball which has been charged by glass and is repelled by it, is attracted to the charged amber rod. In a more direct experiment we find that two suspended charged amber rods repel each other and two suspended charged glass rods repel each other, but a charged glass rod attracts a charged amber rod. These effects can be produced with many materials such as wax, sulfur, plastic, and rubber but every material behaves either as glass or as amber.

We should also note that the materials with which the rods are rubbed are left with charges opposite to those on the rods. For example, the fur which rubbed the amber is charged with glasslike electricity, and the silk which rubbed the glass is charged with amberlike electricity.

These early experiments led to the following two conclusions about electricity

1. There are two kinds of electricity or charge, glasslike and amberlike.
2. Like charges repel each other and unlike charges attract each other.

Benjamin Franklin (1706-1790) introduced the terms we use today. He called glasslike electricity *positive* and amberlike electricity *negative*. This designation arose out of Franklin's concept of electricity. He thought of it as a fluid present in all substances. An object becomes charged if the normal level of the electric fluid is unbalanced. An excess of electric fluid makes the object positive and a deficiency makes it negative. Although we now know that this concept is incorrect, it was very useful at the time.

During this period of early experimentation, two devices were developed which greatly facilitated progress in understanding electricity. In 1672 Otto van Guericke developed a machine for generating a large amount of charge. It consists of a pivoted cylinder made of sulfur in contact with a material such as silk. As the cylinder is rotated, it rubs against the silk and generates charge. The other device is the Leyden jar, invented by Pieter van Musschenbroeck in 1745. It was designed to store a large amount of electric charge and consists of a glass container with metal foils fastened to the inner and outer walls of the jar (Figure 2-2). Metal wires provide contact to the inner and outer foils. The jar is charged by connecting the two wires to the terminals of the generating machine. Charge from the machine is transferred to the foils and is stored there. In this process the two foils become oppositely charged. If the two connecting wires of the charged jar are brought close together, the attraction between the opposite charges forces the charge to jump from one wire to the other. As the charge jumps the gap, it heats the air and produces spark. The opposite charges having come together, the jar is discharged.

Very early in the course of electrical experiments it was found that metals such as gold, copper, and iron conduct charges, whereas materials such as rubber and glass do not. One of many experiments that illustrates this property is shown in Figure 2-3. The two terminals of a charged Leyden jar are connected through a material such as glass. After the glass connection is removed, a spark is still produced when the two metal terminals are brought close together. This shows that the charge did not pass through the glass. On the other hand, if the two terminals of the Leyden jar are connected directly or through a piece of metal, the jar is discharged. Materials that do not conduct electricity are called *insulators*; those that do are called *conductors*.

In 1748 Dr. William Watson (1715-1787), an English physician interested in

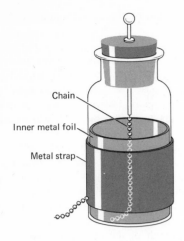

Figure 2-2 Leyden jar. The inside of the glass jar is lined with a metal foil and the outside is surrounded with a metal strap. Electric charge is stored between the two terminals.

electricity, performed an experiment which was very important to the subsequent development of communication. Watson showed that charge can travel long distances. He did this by discharging a Leyden jar through 2 miles of wire. Watson also attempted to measure the time required for the passage of electricity through this long wire and he concluded that the time of passage was shorter than he could measure.

Electricity became a popular game. Large electrostatic generators were constructed and people sent electric sparks through everything they could find. It is reported that in the court of Louis XV, monks from a nearby monastery were formed into a 900-foot circle and a Leyden jar was discharged through them. They all jumped in the air and afterward reported that they were greatly exhilarated by the effect of electricity.

At this point there was enough information about electricity to attempt its use in communication. The first suggestion for an electric communication system was published in the February issue of *Scott's Magazine* in the year 1753 by someone signed C.M. The author proposed a system in which for every letter of the alphabet a wire is strung between the sender and the receiver. At the receiving end, each wire has its own pith ball suspended by a silk thread. At the sending end, charge from a frictional generator is applied to the wire assigned for the letter

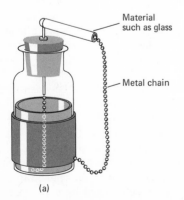

(a)

(b)

Figure 2-3 Experiment with Leyden jar. (a) If a rod made of material such as glass, amber, or rubber connects the two terminals of a charged Leyden jar, the jar remains charged, showing that these materials do not conduct electricity. (b) On the other hand, if the two terminals are connected by metal, the jar is discharged, showing that metal conducts electricity.

to be transmitted. This charge causes a deflection of the corresponding pith ball at the receiving end.

Following this proposal there were many others along similar lines, but clearly there was not the scope for major variations. The differences in the proposed systems were mainly in the methods of coding the alphabet and the ways of detecting the electric signal. In some of the proposed systems the arrival of the signal was to be detected by observation of a spark rather than by the deflection of the pith ball. The system along the lines suggested by C.M. was reported to have been constructed in 1774 by George Louis LeSage (1724-1803) in Geneva. In 1787 M. Lemond in France constructed and operated a pith-ball telegraph which required only one insulated wire for transmission. In the same year Betencour constructed a transmission system using a series of Leyden jars. Dozens of similar systems were constructed around this time, but they were all impractical.

Although the basic principle of these electric telegraphs was correct, the technology was not advanced enough to make them useful. The insulation on the transmission wires was poor, and therefore much of the charge was lost before reaching the receiver. The generation of electricity was by friction and the detection depended on the observation of sparks or deflection of light balls. These methods of charge production and detection were inefficient and cumbersome. It was not until the 1830s that sufficient scientific and technological progress was made to construct a practical electrical communication system.

The main purpose of the following few topics is to give a more detailed explanation of electric phenomena and to lay the foundations for our subsequent discussion of the connection between electricity and magnetism. The understanding of the relationship between electricity and magnetism was the most important step in the development of modern communication systems. On the whole we shall maintain an historical perspective in our discussion of communication, but two topics, "lines of force" and "Bohr atom" are not in historical sequence. These topics are discussed early because they will make the understanding of subsequent topics easier.

COULOMB'S LAW

The first 150 years of experimentation with electricity was mostly qualitative. It was known that a force existed between charged objects, causing them to attract or repel each other, but the dependence of this force on the magnitude of the charge and on the distance between them was not known. The first accurate quantitative measurements of the electric forces between two charged objects were made by Charles A. Coulomb (1736-1806), a French military engineer who turned his attention to basic science. His highly accurate measurements laid the foundation for the quantitative understanding of electricity.

The major step that allowed Coulomb to measure the force between electric charges with great accuracy was his invention of a torsion balance in 1784. (At about the same time the Reverend John Mitchell in England invented a similar instrument which was later used to measure the density of the earth.) The instrument is shown in Figure 2-4. It is basically a simple device consisting of a metal thread at the end of which a rod is suspended. If a force is applied to the end of the rod, the rod and thread twist an amount proportional to the applied force. In describing the torsion balance Coulomb wrote "I determined by experiment the

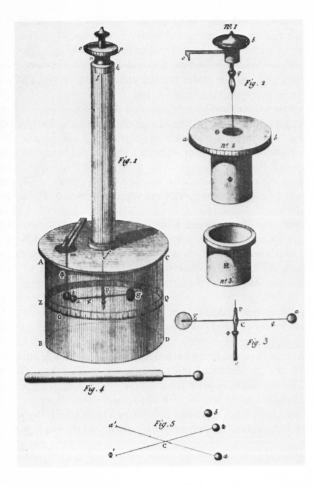

Figure 2-4 Coulomb's drawing of his apparatus. Balls *a* and *t* were charged. The charged balls caused the wire *P-l* to twist. The angle of twist was proportional to the force between the charges.

laws governing the torsion in a metal thread. I found that this force is proportional to the angle of twist, to the fourth power of the diameter, and inversely proportional to the thread's length. The constant of proportionality depends upon the metal used and can be determined experimentally."

With the calibrated torsion balance, Coulomb was able to measure the force between two charged bodies. He placed equal charges on two balls, *a* and *t* (Figure 2-4), and then measured the angle through which the rod twisted as a result of the repulsive force. By changing the amount of charge on *a* and *t* he was able to measure how the force varied as a function of charge. By varying the distance between the balls he determined the force as a function of the distance between two charged bodies. Coulomb also experimented with charges of opposite sign which, of course, attracted each other.

The dependence of the force between two charged bodies on the charge and on the distance between the charged bodies is expressed by the following equation which is now called Coulomb's law:

$$F = K \frac{Q_1 Q_2}{R^2}.$$

Here F is the force between two bodies with charges Q_1 and Q_2 separated by distance R. The constant K depends on the units used in the equation. The quantities Q, F, and R in the equation may be expressed in various units. For example, the distance R may be measured in meters, feet, yards, and so on. In each case a different value of K must be used in order for the equation to be quantitatively valid. This is discussed in greater detail in a later section.

Coulomb's law shows that the electric force between two charged bodies is proportional to the magnitude of the charges. If the charge on one body is doubled, then the force is also doubled. The force is inversely proportional to the square of the distance. Thus, for example, if the distance between the two charged bodies is doubled, the force between them is decreased by a factor of four. Let us note again that the direction of the force depends on the sign of the charges. The force between two like charges is repulsive, whereas the force between unlike charges is attractive.

There is a remarkable similarity between Coulomb's law and Newton's law of gravity. The gravitational force between two masses M_1 and M_2 separated by a distance R is $F = GM_1M_2/R^2$. Here again G is a unit-dependent constant of proportionality. The similarity in the form of the two laws has led many people to search for a connection between gravitational and electric forces. So far such a connection has not been found.

There are, however, some major differences between electric and gravitational effects. In electricity we have two kinds of charges, as a result of which the electric force can be either attractive or repulsive. However, there seems to be only one kind of gravitational mass and the force between these like masses is attractive. Furthermore the forces due to electrical interactions are much larger than the forces due to gravity. This is clearly evident from simple electrostatic experiments. For example, a small charged amber rod or a charged comb can pick up pieces of paper, showing that the electric force between the charged object and the paper is larger than the attractive gravitational force between the earth and the paper.

LINES OF FORCE

An electric charge exerts a force on another electric charge; a mass exerts a force on another mass; and as we will see later a magnet exerts a force on another magnet. All these forces have an important common characteristic: The exertion of the force does not require physical contact between the interacting bodies. The forces act at a distance. A very useful way of visualizing these forces which act at a distance is the concept of *lines of force* or *field lines* which was introduced by Michael Faraday about 50 years after Coulomb's experiments.

Any object that exerts a force on another object without contact can be thought of as having lines of force emanating from it. The complete line configuration is called a *force field*. The lines point in the direction of the force, and their density at any point in space is proportional to the magnitude of the force at that point.

As an example we can consider the lines of force surrounding the earth. The lines point toward the center of the earth since that is the direction of the gravitational force acting on a mass (Figure 2-5). The density of the force lines, that is, the number of lines that intersect a given area, decreases as we move away from the earth. We can express this more quantitatively. The total number of lines converging into the earth is proportional to the mass of the earth M. Since the force

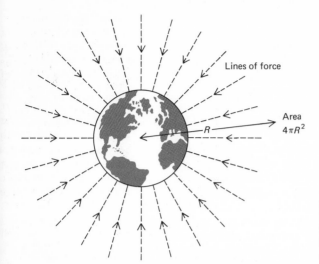

Figure 2-5 Gravitational lines of force surrounding the earth. The density of force lines decreases with distance from the earth.

lines originate and spread out from their source, the number of force lines intersecting a given area decreases as we move away from the source. Specifically at any point in space, a distance R away from the center of the earth, the density of force lines is inversely proportional to the area of a sphere with that radius. Since all the lines of force cross that sphere, the line density is $M/r\pi R^2$ ($4\pi R^2$ is the area of a sphere with radius R). The force exerted on an object with mass m is proportional to the force line density at the point of its location and to its mass. The force on the object m is therefore $Mm/4\pi R^2$. Except for the constant of proportionality, this is Newton's law of gravity.

A somewhat subtle point in connection with the field treatment of forces is that the source of the field, in this case the earth, need not be considered in the computation. In order to compute the force acting on the object with mass m it is sufficient to know the size and direction of the field. Although it is true that for the initial determination of the field the source has to be known, thereafter the field can be treated as an independent entity.

The lines of force associated with electric charges are treated in a similar way (Figure 2-6a,b). The lines of force emanate from the charge uniformly in all directions. By convention the lines point in the direction of the force that the source charge exerts on a positive charge. Thus the lines of force point away from a positive source charge and into a negative source charge. The number of lines emanating from the charge is proportional to the size of the electric charge. If the size of the source charge is doubled, the number of force lines is also doubled. As is the case with the gravitational field the density of force lines decreases as we move away from the source. The density is again inversely proportional to the area of the sphere that surrounds the source charge at the point of interest. The force exerted on another charge Q_2 at a distance R from the source charge Q_1 is proportional to the size of that charge Q_2 and to the density of force lines at the location of Q_2. Therefore the force between two charges Q_1 and Q_2 is proportional to Q_1Q_2/R^2. Except for the constant of proportionality, this is Coulomb's law.

Although the concept of force lines is applicable to all forces acting at a distance, we must remember that the lines act only on entities to which they relate. Thus

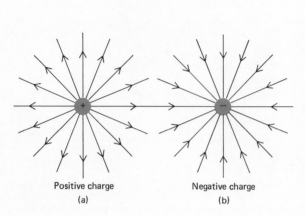

Positive charge
(a)

Negative charge
(b)

Figure 2-6 The diagram shows a two-dimensional representation of the electr lines of force or field lines produced by a positive poi charge (a) and a negative point charge (b). Another positive charge placed in the neighborhood of the source charge experiences a force in the direction of th field lines. The magnitude (the force is proportional to the density of the lines at th location of the charge. The density of the lines de-creases as the distance fro the source increases. In an actual three-dimensional situation, the field density a a point a distance R from th source is inversely propor-tional to the area of a spher of radius R which is $4\pi R^2$. Therefore the force betwee two charges is proportional to $1/R^2$.

electric lines exert forces only on electrically charged objects and gravitation lines exert forces only on masses.

Lines of force need not be straight lines; as we mentioned they point in the dire tion in which the force is exerted. As an example, we can consider the net field due two charges separated by a distance d. To determine this field we must comput the direction and size of the net force on a positive charge at all points in space. Th is done by adding vectorially the force lines due to each charge. The force fiel due to a positive and negative charge separated by a distance d from each other shown in Figure 2-7. Here the lines of force are curved. This is, of course, th direction of the net force on a positive charge in the region surrounding the tw fixed charges. The field shown in Figure 2-7 is called a *dipole* field and, as we wi see, it is similar to the field produced by a bar magnet.

THE BOHR ATOM

One of the most important contributions to modern physics has been the model the atom proposed in 1913 by the Danish physicist Niels Bohr (1885-1962). Thi model showed the way out of the impasse which atomic physics had reached a that time. Not only did it explain many of the experimental results that were bafflin physicists, it also opened the way for the development of modern physics. W now know that the Bohr model is conceptually not completely correct, but it adequate for the explanation of the phenomena we will encounter in this discussior In explaining the Bohr atom we adopt a spiral approach. We discuss it a numbe of times throughout the text, each time in greater detail as the surrounding discussion requires it. In this, our first encounter with the model, we will give enough of its basic features to explain electricity and magnetism.

In 1911 Niels Bohr, having just received a Ph.D. in Copenhagen, went to England for further study at Cambridge University. England at that time was the center for atomic research, with the two most important research groups being led by J. J. Thompson (1856-1940) and E. Rutherford (1871-1937). A number of important facts had already been gathered about atoms which make up matter. For example, it was known that atoms contain small negatively charged *electrons* and relatively heavier positively charged *protons*. The proton is about a thousand times heavier than the electron, but the magnitude of the charge on the two is the same. There are as many positively charged protons in an atom as negatively charged electrons. The atom as a whole is therefore electrically neutral. The identity of an atom is determined by the number of protons it has. For example, hydrogen has 1 proton, carbon has 6 protons, silver has 47 protons. Through a series of ingenious experiments Rutherford showed that most of the atomic mass is concentrated in a nucleus made up of the protons. The electrons therefore have to be situated in some way outside of the nucleus. It was known that the size of the atom is on the order of 10^{-8} centimeters[1] (0.00000001 centimeter) and Rutherford showed that the nucleus is on the order of 10^{-13} centimeter. Thus the nucleus is very dense and most of the atom is void.

When Bohr first became acquainted with atomic physics, the subject was in a state of confusion. A number of theories had been proposed for the structure of the atom, but none explained satisfactorily the existing experimental results. The most peculiar observed property of atoms was the light emitted by them. When an element is put into a flame, it emits a series of characteristic colors, called *spectral*

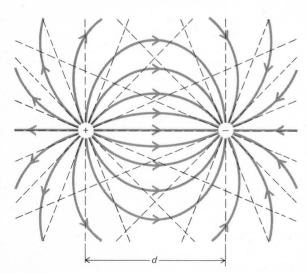

Figure 2-7 Lines of force produced by two charges add vectorially. The broken lines show the field due to the individual charges. The solid lines show the resultant field due to a positive and a negative charge fixed at a distance *d* from each other. A positive test charge would move along the lines of force in the direction of the arrows. A negative test charge would move in the direction opposite to the arrows. This field configuration is called a *dipole field*.

[1]The notation used above is convenient for expressing large and small numbers. The following illustrates the use of this notation.

$1 = 10^0$	$1000 = 10 \times 10 \times 10 = 10^3$	$0.001 = 1/1000 = 10^{-3}$
$10 = 10^1$	$0.1 = 1/10 = 10^{-1}$	$2320 = 2.32 \times 10^3$
$100 = 10 \times 10 = 10^2$	$0.01 = 1/100 = 10^{-2}$	$0.00232 = 2.32 \times 10^{-3}$

lines. Prior to Bohr, scientists could not explain why these colors were emitted by atoms.

After he returned to Copenhagen, Bohr proposed a model for the atom that not only explained the reason for the spectra but also made some predictions about new spectral lines in hydrogen which were later observed. Bohr started with the model of the atom as proposed by Rutherford. At the center of the atom is the positive nucleus made up of protons. (It was subsequently discovered that the nucleus also contains another particle, the neutron, which has approximately the same mass as the proton but is electrically neutral.) The electrons orbit around the nucleus much as the planets orbit around the sun. They are maintained in orbit by the electrostatic attraction of the nucleus. And here is the major feature of the Bohr model. In order for the model to explain the emission of spectral lines, it was necessary to postulate that the electrons are restricted to distinct orbits around the nucleus. In other words, the electron can be found only in certain allowed orbits. Bohr was able to calculate the radii of these allowed orbits and show how the spectral lines are emitted as a consequence of the orbital restrictions. Bohr's calculations are found in most elementary physics texts (see References).

The orbital restrictions are most easily illustrated with the simplest atom, hydrogen, which has a single-proton nucleus and one electron orbiting around it (Figure 2-8). Unless energy is added to the atom, the electron is found in the allowed orbit closest to the nucleus. If energy is added to the atom, the electron may "jump" to one of the higher allowed orbits further away from the nucleus, but the electron can never occupy the regions between the allowed orbits. When energy is added to atoms, the electrons are transferred to higher allowed orbits. As they return to their original lower orbit the energy is released as light, which makes up the observed characteristic spectral lines.

The Bohr model was very successful in explaining many of the experimental observations for the simple hydrogen atom. But in order to describe the behavior of atoms with more than one electron it was necessary to impose an additional restriction on the structure of the atom: The number of electrons in orbit n cannot be larger than $2n^2$. Thus the maximum number of electrons in orbit 1 is $2 \times (1)^2 = 2$, in orbit 2 is $2(2)^2 = 8$, in orbit 3 is $2(3)^2 = 18$, and so on. The atoms are found to be constructed in accord with these restrictions (Figure 2-9). Helium has two electrons and therefore its first orbit is filled. Lithium has three electrons, two of which fill the first orbit; the third electron therefore must be in the second orbit. Of course, for each electron in orbit there is a proton in the nucleus, thus making the atom electrically neutral. The second orbit has a capacity for eight electrons and as such

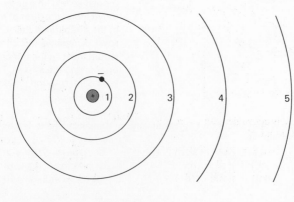

Figure 2-8 Bohr model for the hydrogen atom. The electron orbits about the nucleus and can occupy only discrete orbits with radii 1, 2, 3, and so on. In the absence of external perturbations the electron is found in the smallest allowed orbit. If energy is added to the atom, the electron may "jump" to one of the higher orbits.

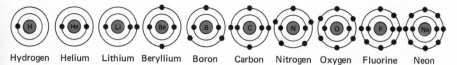

Hydrogen   Helium   Lithium   Beryllium   Boron   Carbon   Nitrogen   Oxygen   Fluorine   Neon

Figure 2-9   The Bohr model for the first 10 atoms in the periodic table. Note that the orbits are filled in accord with the discussion in the text.

cessive electrons are added to this orbit, the elements beryllium, boron, carbon, nitrogen, oxygen, fluorine, and neon are formed. With neon the second orbit is filled and the next element, sodium, has its additional electron in the third orbit. This simple sequence is not completely applicable to the very complex atoms, but basically this is the way all the elements are constructed.

Using this model for the atom it is possible to explain, at least qualitatively, many properties of matter. For example, the formation of chemical compounds and matter in bulk is due to the distribution of electrons in the atomic orbits. When an orbit is not filled to capacity (which is the case with most atoms), the electrons of one atom can partially occupy the orbit of another. This sharing of orbits draws the atoms together and causes bonding between atoms. As an example we show in Figure 2-10 the formation of a hydrogen molecule from two hydrogen atoms. In the orbit of each of the hydrogen atoms there is room for another electron. A completely filled orbit is the most stable configuration; therefore when two hydrogen atoms are close together, they share each other's electrons and in this way the orbit of each atom is completely filled part of the time. This shared orbit can be pictured as a rubber band pulling the two atoms together. Therefore the sharing of the electrons binds the atoms into a molecule. While the sharing of electrons pulls the atoms together, the Coulomb repulsion of the nuclei tends to keep them apart. The equilibrium separation between atoms in a molecule is determined by these two counter forces. In a similar way more complex molecules are formed and ultimately bulk matter.

Atoms with completely filled orbits (these are atoms of the so-called noble gases — helium, neon, argon, krypton, and xenon) cannot share electrons with other elements and are therefore chemically inert.

By adding sufficient energy to the atom it is possible to rip off one or more of its electrons. Such an atom with missing electrons is, of course, positively charged and is called an *ion*.

Although the Bohr model explained many observations, from the very beginning the theory appeared contrived and raised more questions than it answered. The model, however, was a daring step in a new direction which eventually led to the development of quantum mechanics. In the quantum mechanical description of the atom it is not possible to assign exact orbits or trajectories to electrons. Elec-

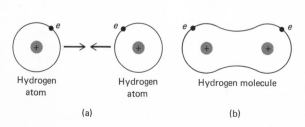

Hydrogen atom   Hydrogen atom   Hydrogen molecule

(a)   (b)

Figure 2-10   A schematic representation for the formation of a hydrogen molecule. (a) Two separate hydrogen atoms. (b) When the two atoms are close together, the electrons share each other's orbit which results in the binding of the two atoms into a molecule.

trons possess wavelike properties and behave as clouds around the nucleus. Th
artificial postulates in Bohr's theory are a natural consequence of the quantum
mechanical approach to the atom. For further discussion of quantum mechanics th
reader is referred to the texts listed in the References at the back of the book.

EXPLANATION OF THE ELECTRIC EXPERIMENTS

We can now explain the results of the electric experiments described earlier. Ob
jects can be electrically charged by rubbing because when two objects are brough
into close contact, the surface electrons from one object enter the surface of th
other. When the two objects are rubbed against each other, electrons are removed
from one object and deposited on the other. In this way when amber is rubbed b
fur, electrons are removed from the fur and are deposited on the amber. Amber
with excess electrons, therefore becomes negatively charged and since those elec
trons have been removed from the fur, it is left with an excess positive charge.

The charged amber initially attracts a pith ball and other small objects that hav
no charge. This is explained by the initial redistribution of charges in the uncharge
body which are caused by the proximity of the charged object. For example, in th
case of the amber and the pith ball, the negatively charged amber pushes some c
the electrons away from the pith-ball surface. This leaves a net positive charge o
the pith-ball surface nearest to the amber (Figure 2-11). As a result of the forc
between the negative amber and the exposed positive charges on the ball, the tw
objects are attracted to each other. When contact is made between the pith ba
and the rod, some of the excess electrons on the rod are transferred to the pith ba
and also make it negative. Now both objects are negatively charged and they repe
one another. The behavior of glass when it is rubbed by silk is similarly explaine
except that here the electrons are removed from the glass rod and deposited on silk.

We have noted that some materials conduct electricity, whereas others insulate
This can be explained at least qualitatively by the type of interatomic bonding tha
exists in the material. In conducting materials, such as copper or silver, some c

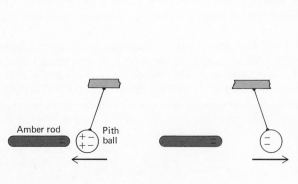

Amber rod   Pith
                   ball

Figure 2-11 Electric exper
ment with a pith ball. (a)
Before contact with the
charged amber rod, the pith
ball is neutral but the
charges on it are distributed
in such a way that it is attracted by the amber. (b)
When the pith ball makes
contact with the charged
amber, some of the excess
electrons are transferred
from the amber to the pith
ball. Both objects are now
negatively charged and
therefore repel each other.
The experiment works more
efficiently if the pith ball is
coated with a metal foil or
conducting paint which aids
the redistribution of charge

the electrons are held rather loosely to the nuclei. These electrons do not specifi-cally belong to any single atom but travel freely throughout the material (Figure 2-12. When we place excess electrons on one end of a conductor, these charges can now also freely move within the material and all the charges redistribute themselves throughout the material in such a way as to minimize their mutual repulsions. If we continue to place negative charges at one end and remove them from the other end, the redistribution becomes continuous, resulting in a flow of charges from one end of the material to the other.

In insulating materials, such as rubber or sulfur, the electrons are tightly bound and localized. They are not free to move within the material. If an excess charge is placed at one end of an insulator, the orbits of the electrons are altered, but no excess charge appears at the other end of the material. The charges cannot travel through the material.

NTRODUCTION TO MAGNETISM

Prior to the 1600s the understanding of magnetism was on about the same level as the understanding of electricity. In fact, not much distinction was made between the two. The Greeks discovered that a certain kind of iron oxide (called *magnetite* or *lodestone*) attracted or repelled other pieces of the same material or small bits of iron. When a piece of magnetic material was freely suspended, one end of it pointed North (Figure 2-13). It is not known exactly when this effect was discovered, but it was used as a navigational aid by the twelfth and thirteenth centuries.

Many of the experimenters who did the early work on electricity also studied magnetism. They found many similarities between electric and magnetic phenomena. Just as in electricity, in magnetism there also are two kinds of entities, the north and south poles, and both attraction and repulsion can be observed between mag-netic materials. It is again observed that like poles attract and unlike poles repel each other (Figure 2-14). There are, however, some important differences between electricity and magnetism. It was William Gilbert around the year 1600 who made the clear distinction between magnetism and electricity. Although positive and negative electric charges can exist separately, the north and south magnetic poles cannot be separated. If a magnetic bar is broken, the opposite north and south poles appear at the point of fracture (Figure 2-15). A charged rod can be neutralized by touching it with a grounded conductor, but a piece of magnetite will retain magne-tism essentially indefinitely. Magnetism is not conductible the way electricity is.

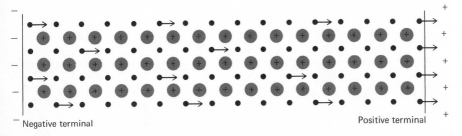

Negative terminal   Positive terminal

Figure 2-12 Conduction of electricity in a metal due to the motion of the electrons from the negative to the positive terminal. Electrons in a metal are bound weakly to the surrounding atoms and can therefore move within the metal.

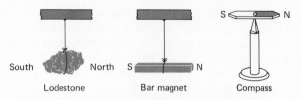

South   North   S ▬▬▬▬▬ N

Lodestone   Bar magnet   Compass

Figure 2-13 Magnetized objects which are free to turn align in a north-south direction.

The lines of force surrounding a magnet are shown in Figure 2-16. The field pattern is similar to that produced by a positive and negative charge separated by a fixed distance. This is not surprising since the magnetic field is in fact produced by the two opposite entities, the north and south poles. The pattern of the field can be demonstrated by placing small magnets or iron filings in the neighborhood of the magnet. These small magnetic objects will line up along the direction of the lines of force, but they will not move along these lines because they themselves have a north and a south pole. The forces acting on the two poles are equal and opposite thus canceling each other. Of course, if a magnetic pole could be found separated from its opposite, it would move along the lines of force. There is, in fact, at present no theoretical reason why magnetic poles could not exist separately, but all attempts to find them have failed so far. The search will go on, however, until either "mono poles" are found or the reason for their nonexistence is understood.

We have seen how electrostatic phenomena alone could be, at least in principle used for communication. There is no practical way of communicating by the use of magnets. The idea, however, was attractive and as early as 1617 an Italian Jesuit priest, Famianus Strada (1572-1649) suggested communication with magnets. He proposed that if a magnetic needle were moved on a dial marked by the alphabet, another needle an arbitrary distance away would move to the same letter. Although it was shown that magnets do not behave in this way, suggestions of this type were being made for many years.

Through Gilbert's work it was known that electric and magnetic phenomena are different. Still most experimenters, including Gilbert, suspected that there was a connection between them. Attempts were made to establish the relationship between the two effects. Charged objects were brought near magnets, but no mutual force was observed between them. The connection between electricity and magnetism was not established until 1820. In retrospect it is not surprising that this took so long. Stationary charges do not affect magnets. In order to interact with a magnet a charge has to be in motion. Charges in motion are called *electric current*. In the electricity experiments we have discussed up to now, the charges were generated by friction and they were mostly stationary. A sudden motion of charges could be

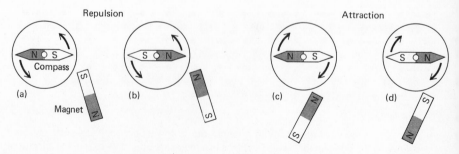

Figure 2-14 Like magnetic poles attract each other, unlike magnetic poles repel each other

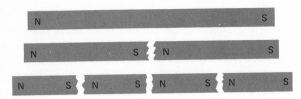

Figure 2-15 Magnetic poles do not exist alone. When a magnet is broken, opposite poles appear at the broken ends.

produced by discharging a Leyden jar or some other object; but this current was small and of short duration. A steady source of electric current was not available until the 1790s when Alessandro Volta invented the electrochemical cell.

LECTROCHEMICAL CELL AND BATTERY

The electrochemical effect was first observed in 1780 by Luigi Galvani (1737-1798), who was a physician and professor of anatomy at the University of Bologne. The discovery was quite accidental. It is reported that one of Galvani's assistants happened to touch the nerve of an amputated frog leg with a metal scalpel at the moment when an electrostatic machine nearby made a spark. The leg gave a small kick. The coincidence of the kick with the electric spark led Galvani to conclude that the phenomenon had something to do with electricity. He soon discovered, however, that the spark was not necessary. The convulsion was produced every time the frog's leg was laid on a metal surface and the nerve was touched by a metal scalpel, which at the same time also made contact with the metal surface. Galvani continued to believe, correctly so, that the effect was connected with electricity, but he concluded wrongly that the source of electricity was in the frog's leg. He published the results of his experiment in 1791 and called the vital force in the tissue which caused the motion of the leg *animal electricity*.

At the time of Galvani's discovery Alessandro Volta (1745-1827) was already working with electricity and had discovered a number of interesting devices. Volta became very interested in Galvani's results and for a while he also believed that the effect was caused by animal electricity. However, as Volta continued to experiment he became convinced that the phenomenon was not that of animal electricity but that the frog's leg simply responded to a current generated by two dissimilar metals. He found that when two dissimilar metal rods such as zinc and copper, or zinc and silver, are placed in slightly acidic or salty water, electric charges are produced at the two electrodes. In the 1790s when Volta announced his results, there were very few who believed him. His arguments with Galvani were especially bitter. The

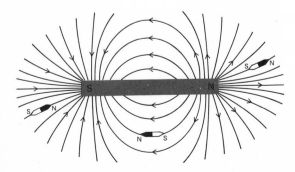

Figure 2-16 Magnetic lines of force around a straight bar magnet. The direction of the field can be traced by observing the alignment of small magnets placed in the proximity of the bar magnet.

amount of charge produced by the type of cell we described is very small ar
therefore Volta was not able to produce a spark or any other effect then conve
tionally associated with electricity. He spent two years in an effort to enlarge t
effect of his electrochemical cell. He finally hit upon a way of connecting the ce
so that the charges of individual cells were added. He was then able to demonstra
that the electricity produced by his cells was the same sort as the electricity produce
by friction. There was, however, an important difference. The voltaic cell was
continuous source of charge. A steady current, large enough to produce a powerf
arc for many minutes, was provided by the voltaic battery.

It was a great triumph for Volta to prove that the dissimilar metals were th
source of electricity. In 1801 Volta demonstrated his battery to an importai
audience which included Napoleon. He took great care to make his battery appea
like an electric eel, probably to mock the idea of animal electricity so strongl
defended by Galvani. Volta did not understand the operation of his cell. He wrot
"It may seem paradoxical and may prove inexplicable but it is nonetheless real.
Although the operation of the cell was not explained until 1864, it was nevertheles
extensively used and brought about the development of electromagnetism.

The operation of the voltaic cell is actually easy to explain in terms of ou
present understanding of matter (Figure 2-17). When the zinc rod is placed in wate
some of the zinc atoms from the surface of the rod dissolve into the water as ion:
In the case of zinc, for each ion that dissolves in water two excess electrons ar
left on the rod. The process of dissolving continues until there is a specific concen
tration of ions in the solution. Since the rod is negative, it attracts the positiv
ions back to the surface at the same time as ions are leaving it. In equilibrium a
many ions leave the surface as are deposited on it, but in the process the rod accumu
lates a negative charge. The identical process occurs with other metals, such a
silver or copper, when they are immersed in water. The amount of negative zin
ions obtained from a zinc rod in water is higher than the concentration of coppe
ions obtained from a copper rod. Therefore the zinc rod is left with greater negativ
charge than the copper rod. If a zinc and a copper rod are immersed in the same con
tainer, both rods will have an excess amount of electrons with respect to the solu
tion, but there will be a larger excess of electrons on the zinc rod than on the coppe
rod. When the two rods are externally connected by a conducting material, the
electrons move from the zinc rod, which is a region of higher charge density, to the

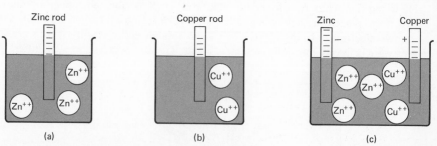

Figure 2-17   Voltaic cell. (a) If a zinc rod is immersed in water, positive zinc ions dissolve ir
water leaving the rod with an excess negative charge. (b) The same process occurs with
a copper rod, but the copper rod is less negative than the zinc rod. If the zinc and the
copper rods are immersed in the same solution, a voltaic cell is formed. (c) Since there
are more electrons on the zinc rod than on the copper rod, the zinc rod is negative with
respect to the copper rod.

copper rod. This current is maintained by the continuing activity in the solution. As the electrons leave the zinc rod the rod becomes less negative and therefore more zinc ions can leave it. On the other hand, the copper rod becomes more negative and therefore more copper ions are attracted to it. The current continues until the zinc rod becomes corroded. The cell works much more efficiently if instead of pure water, a dilute acid or salt solution is used. If these individual cells are connected as shown in Figure 2-18, their effects become cumulative. Such a configuration is called a *voltaic battery.*

Soon after the discovery of the electrochemical cell, proposals were made to use it for communication. On May 2, 1800, Anthony Carlisle (1769-1840) and William Nicholson (1753-1815) showed that when two electrodes connected to a voltaic pile were immersed in water, some of the water decomposed forming bubbles of oxygen at one electrode and bubbles of hydrogen at the other. This observation led in 1809 to a proposal of an electrochemical telegraph by the German anatomist Samuel Thomas von Sömmering (1755-1830). Sömmering's system was impractical, but it does illustrate the close coupling between the discovery of an effect and the attempt to apply it. The apparatus is shown in Figure 2-19. At the receiving end 35 pins marked by the 25 letters of the German alphabet and the 10 numerals were immersed in acidulated water. Each pin was connected by wire to the transmitting station, where current was supplied to a pair of wires from the voltaic battery. This current decomposed the water at the receiving end and produced bubbles at the pins corresponding to the two wires. Hydrogen was produced at one wire and oxygen at the other. Because a water molecule is made of two hydrogen atoms and one oxygen atom, more hydrogen was emitted than oxygen. The operator read the letter from the pin with the larger emission of gas. In order to alert the operator that a message was coming, an alarm system was devised. A beaker was inverted over an additional pair of pins in the solution. The beaker was connected by a set of levers to an alarm clock. The alerting signal produced bubbles under the beaker, causing it to move and activate the alarm clock.

MAGNETIC FIELD FROM AN ELECTRIC CURRENT

With the invention of the battery the stage was set for the discovery of the long sought connection between electricity and magnetism. The first half of the connection was found in 1820 by a Danish scientist, Hans Christian Oersted (1777-1851).

Oersted's great discovery was reportedly accidental. As the story is usually told, he was lecturing to his students on electricity and magnetism. As part of the

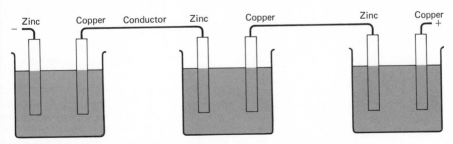

Zinc   Copper   Conductor   Zinc   Copper   Zinc   Copper

Figure 2-18 Voltaic battery. When voltaic cells are connected as shown, their effect becomes cumulative.

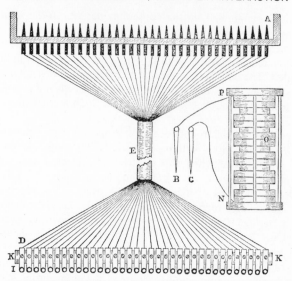

Figure 2-19 Diagram of Sömmering's proposed electrochemical telegraph. (Courtesy of the New York Public Library.)

demonstration he connected the two terminals of the voltaic cell to produce a current flow. By chance there was a magnetic compass needle near the wire, and he observed that the needle rotated to a position at right angle to the current-carrying wire. When he reversed the connections to the battery (thus reversing the motion of the charges) the needle reversed its direction but remained perpendicular to the wire.

Oersted realized the importance of this effect: He had found the connection between electricity and magnetism. It is the motion of charges that causes the attraction of the magnet. Oersted found that the compass needle was always aligned perpendicular to the current-carrying conductor (Figure 2-20). He concluded therefore that charges in motion produce a concentric magnetic field around the conductor and the magnet is aligned by this field. When the current is reversed, the field is reversed with it. The current-carrying conductor behaves as a magnet, but its magnetic field distribution is different from that of a permanent magnet. The magnetic field produced by a single current-carrying conductor has no poles.

At this point we must explain the convention adopted for the direction of current flow. The designation of current direction started long before it was known that in a

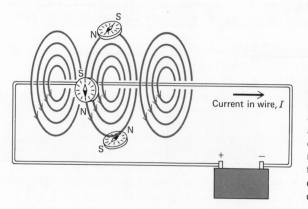

Current in wire, I

Figure 2-20 Oersted's discovery. A current-carrying conductor exerts a force on a magnet. This implies that a current-carrying conductor is surrounded by a magnetic field. The direction of the magnetic field is determined by the direction of the current flow.

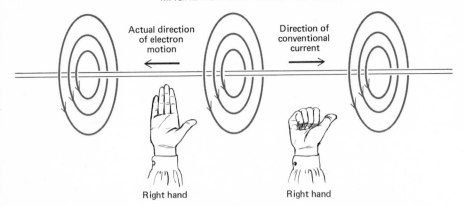

Actual direction of electron motion

Direction of conventional current

Right hand   Right hand

Figure 2-21   The right-hand rule. The direction of the magnetic field surrounding a straight current-carrying wire is obtained by the application of the "right-hand" rule. With the thumb of the right hand pointing in the direction of the conventional current flow, the fingers point in the direction of the magnetic field.

conductor the current is carried by electrons. It was then arbitrarily decided that the current flows from the positive to the negative terminal of a battery. We now know that the electrons flow from the negative terminal to the positive. However the old conventional current designation is still frequently used.

As we mentioned, the direction of the magnetic field is determined by the direction of the current that produces it. There is a useful mnemonic called the *right-hand rule* which specifies the direction of the field (Figure 2-21). With the thumb of the right hand pointing in the direction of conventional current (opposite to the electron flow), the fingers point in the direction of the magnetic field.

Oersted continued his experiments and published the results in 1820. The importance of his discovery was recognized immediately and it gave a great impetus to further study of electromagnetic effects. One of the people who became very interested in this field was André Marie Ampère (1775-1836). He extended Oersted's experiments to show that two current-carrying conductors exert forces on each other. Although there are moving charges in the conductor, the conductor is not charged. The moving electrons are neutralized by the positive charges through which they move. The force between the conductors therefore is not electric, it is magnetic. The magnetic field configuration between two conductors carrying current in the same direction is similar to the field between the south and north poles of two bar magnets (Figure 2-22). We know that the force between the south and north poles of two magnets is attractive; therefore the force between two conductors carrying currents in the same direction is also attractive. If the conductors carry currents in the opposite directions, the force between them is repulsive.

Ampère was a mathematician and was able to devise a mathematical description for these interactions which was most useful in further work. He published these findings in 1822 and 1823. Ampère foresaw the possible use of Oersted's effect in communication. He suggested that a current-carrying conductor can control the motion of a magnetic needle to spell out alphabetical letters. There were many others who made great contributions to this field immediately after Oersted's discovery. Prominent among them were Jean Baptiste Biot (1774-1862), Félix Savart (1791-1841), Sir Humphry Davy (1778-1829), and Michael Faraday (1791-1867). The work of some of these people will be discussed later.

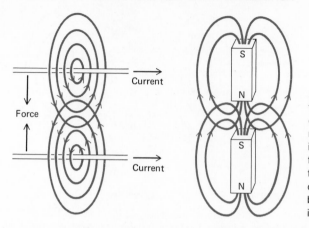

Figure 2-22   The magnetic field in the proximity of tw conductors carrying a current in the same direction is similar to the field configuration produced betwee the north and south poles of two magnets. The force between the two conducto is therefore attractive.

The magnetic field around a straight current-carrying conductor bears little re semblance to the field of a permanent magnet. However, if the wire is formed int a loop, the field pattern begins to resemble the field of a small magnet (Figure 2-23 In 1820 a German chemist, Johann S. C. Schweigger (1779-1857), observed that a an insulated conductor is wound into many turns, the fields of the individual turn accumulate and the overall magnetic field becomes much stronger. He called thi device a galvanomagnetic multiplier (Figure 2-24). The field pattern of the coil i identical to that of a bar magnet. A year later James Cumming, a Scotch physicis at Cambridge University, constructed a similar device. He realized that the coi could detect very feeble electric currents by causing a magnetic needle to turn an therefore it could be used in electric telegraphy. Shortly after this it was found tha the magnetic field is greatly increased by placing a soft iron core inside the coil This was an invention of great technological importance which brought about th development of many devices.

Two instruments based on these discoveries and used for the detection o current are the *galvanometer* and the *ammeter*. We saw earlier that there is a force between a current-carrying coil and a magnet. Now if the magnet is held fixed, the current-carrying wire will move. This is the principle of operation for both the

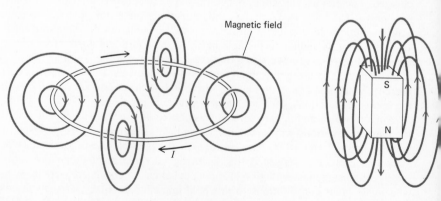

Figure 2-23   The magnetic field pattern produced by a current loop is similar to the field of a short magnet.

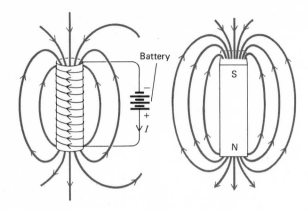

Figure 2-24 Magnetic field produced by a coil is similar to that produced by a long magnet.

galvanometer and the ammeter (Figures 2-25 and 2-26). A coil wound around a soft iron core is placed between the north and south poles of a fixed horseshoe permanent magnet. When a current is passed through the coil, a magnetic field is produced which is proportional to the current. The resulting force between the field produced by the current and the field of the permanent magnet tends to align the coil with the north and south poles of the magnet. A spring restrains the motion of the coil and therefore the coil rotates to a position where the magnetic force is balanced by the restoring force of the twisted spring. The degree of rotation by the coil is proportional to the magnitude of the current passing through it. The difference between a galvanometer and an ammeter is in the suspension of the coil and the detection of the coil rotation. In the galvanometer the coil is suspended by metallic ribbons which also supply the restoring force. The deflection of the coil is measured by the displacement of a beam of light reflected from a mirror attached to the coil. The instrument is delicate but very sensitive. In the ammeter the coil is pivoted between the poles and the restoring force is provided by a coiled spring. The rotation of the coil is indicated by the pointer attached to the coil. The ammeter is less sensitive but more rugged than the galvanometer.

There are many other devices such as motors, relays, and loudspeakers whose operation is due to the force between a current-carrying conductor and a magnet. We will discuss some of these later.

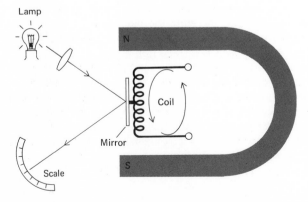

Figure 2-25 Galvanometer. A current passing through the coil causes it to twist in the field of the permanent magnet. The deflection of the coil and the attached mirror is proportional to the current. The deflection is measured by observing the position of the reflected light on the scale.

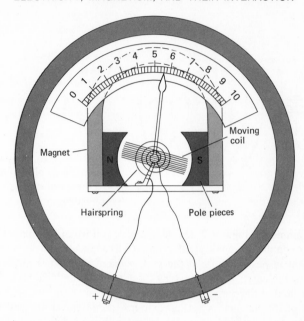

Figure 2-26 Ammeter. A current through the coil produces a magnetic field which causes the coil and the attached pointer to twis in the field of the permaner magnet. The position of the pointer is determined by th size of the current.

THE NATURE OF MAGNETIC MATERIALS

In terms of Oersted's discovery it is possible, at least qualitatively, to explain th magnetic nature of some materials. In the Bohr model of the atom, electrons circl the nucleus and therefore each electron moving around the nucleus can be though of as a small current loop with an associated magnetic field. In some atoms th motion and configuration of electrons are such that the magnetic fields of th individual electrons cancel and the atom as a whole is not magnetic. In other atom such as iron, cobalt, and nickel, the cancelation is not complete and the ato behaves as a small magnet. In materials such as magnetite, the smaller atomic mag nets couple very strongly. Their magnetic fields add and as a result the material a a whole is a magnet (Figure 2-27a). The situation is somewhat different in soft iro The iron atoms have magnetic fields, but the coupling between the atoms occur only over small regions called *magnetic domains*. These small domains in the bul material are randomly oriented and therefore the bulk material has no magneti field (Figure 2-27b). In the presence of an external magnetic field the domains ar aligned in the direction of the applied field. The iron becomes magnetized and th fields of the individual domains are added to the applied field. It is for this reaso that the magnetic field of a conducting coil is made stronger by the insertion of a iron bar. It was Ampère who first suggested that the magnetic properties of matte are caused by permanent currents in the material; however, he was not able postulate a satisfactory source for these currents.

It is now known that the electrons themselves behave as small magnets and tha it is the magnetism of the electrons which accounts for most of the magnetic proper ties of materials such as iron. However, this property of electrons was not know until the 1920s.

LECTROMAGNETIC INDUCTION

We have seen that a current produces a magnetic field. It is therefore logical to ask if a magnetic field can produce a current. This question was answered by Michael Faraday in 1831. He showed that a moving magnet generates a current in a conductor. The current flows as long as there is relative motion between the magnet and the conductor. When the motion stops, so does the current.

When Faraday discovered electromagnetic induction, he was 40 years old and was already a well-known scientist. He had discovered benzene, electrolysis, and had performed very important experiments in optics, chemistry, and metallurgy. With his discoveries in electromagnetism he became the best-known scientist of his time. His accomplishments are even more remarkable in view of his background. Faraday was born on September 22, 1791, near London. His father was a blacksmith and was too poor to keep him in school. At the age of 13 knowing very little besides reading and writing, Faraday took a job as errand boy in a bookshop. A year later he became a bookbinder apprentice for seven years. During this time he read almost anything he got his hands on, mostly books that were being bound in the shop. He was especially fascinated by scientific writings. After acquiring some book knowledge of science, he attended a series of lectures on chemistry given by the great scientist Sir Humphry Davy. Faraday was very impressed; he took detailed notes at the lectures and immediately applied for a job at the Royal Society. He was rejected.

After his apprenticeship, Faraday took a job as a bookbinder but was not satisfied with the work. He applied to Davy for employment, and submitted his lecture notes as proof of his serious intentions. Davy, who was a vain man, hired Faraday as his secretary, but a few months later fired him. A short time later Davy rehired Faraday, this time as his laboratory assistant.

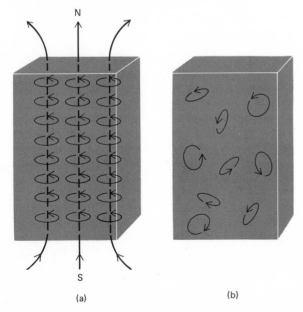

(a)

(b)

Figure 2-27 Magnetism. (a) In a permanent magnet the individual magnetic domains are aligned and therefore their magnetic fields add. (b) In soft iron the domains are randomly oriented and therefore do not give rise to a net magnetic field.

For the next few years Faraday did work in chemistry, metallurgy, and optic all quite successfully. In 1820 when Oersted announced his discovery, Farada turned his attention to electricity. He duplicated and extended Oersted's resul and, shortly afterward, discovered the principle of the electric motor. He wa nominated for membership in the Royal Society and although Davy voted again him, he was elected in 1824.

Faraday felt intuitively that since currents produced magnets, then magnet should produce currents. He unsuccessfully tried to produce a current by placin wire next to a magnet. He then abandoned his experiments in electricity and fc the next six years worked mostly in chemistry. In 1831 he resumed his electric research and on August 29, 1831, he was successful in producing a current with magnet. But he was somewhat disappointed because a current was induced onl when the magnet was moving. To a friend he wrote, "I am busy just now again o electromagnetism and I think I have got hold of a good thing, but can't say. It ma be a weed instead of a fish that, after all my labor, I may at last have pulled up. During the next few weeks Faraday extended his experiments and completed th phenomenological explanation of the process.

Faraday did not have sufficient background to approach the subject mathe matically. Therefore, in order to understand electromagnetic induction, he postulate a physical model, the lines of force. As we have mentioned before, this concept wa most important and useful in the development of physics. He wrote, "All I can sa is that I do not perceive in any part of space, whether vacant or filled with matter anything but forces and the lines in which they are exerted," and "In this view o the magnet, the medium or space around it is as essential as the magnet itself being a part of the true and complete magnetic system."

Although Faraday invented the most useful electric devices—the motor, trans former, and generator—he was not interested in their practical applications. He continued his work in pure research. The nature of gravity intrigued him. He be lieved that there is a connection between electricity and gravity, but his experi ments failed to show any relationship. "Here end my trials for the present," he wrote "The results are negative. They do not shake my strong feeling of the existence o a relation between gravity and electricity." To this day the relationship has no been found.

Faraday conducted the most systematic investigation of the currents induced by changing magnetic fields. There were two others, however, Joseph Henry i the United States and Heinrich Lenz in Russia, who simultaneously and inde pendently studied the same phenomena. Since both the United States and Russia were at the time at the periphery of scientific activities, their work did not receive immediate attention. Unlike Faraday, Henry was interested in the practical appli cations of electromagnetism and after he became director of the Smithsonia Institution, he helped to bring about the development of the electric telegraph

Faraday's experiment illustrating electromagnetic induction is shown in Figure 2-28. When he moved the magnet into the coil, a current was induced in the coil When the magnet was withdrawn from the coil, the current flowed in the opposite direction. Faraday found that the current is directly proportional to the strength of the magnetic field and the speed with which the magnet is moved in the coil. He realized that a current is induced in a conductor when it is in a changing magnetic field.

Physical motion between the wire and magnet is one way of producing a changing magnetic field. There is, however, a method of producing a changing magnetic field

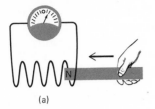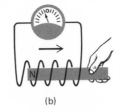

Figure 2-28 Faraday's experiment. (a) A current is
induced in the wire only
while the magnet is in motion. (b) Reversing the
motion of the magnet re-
verses the current.

(a)   (b)

without motion. Consider the experiment illustrated in Figure 2-29. When the switch is closed, a current from the battery flows through coil A (primary coil). Naturally this current produces a magnetic field that extends to coil B (secondary coil). With a continuous current flowing through coil A, the magnetic field is not changing and there is no current flowing in coil B. But when the switch is opened, the current stops flowing and therefore the magnetic field changes from its previous value to zero. During this short period a current flows through coil B, as is indicated by the ammeter. Similarly, if the switch is closed, the magnetic field builds up from zero to a finite value, and during the time of build-up a current is again induced in the coil B. The current in this case is in the opposite direction from that observed during the turn off of the switch. This experiment was one of a series that Faraday performed to establish the nature of induction. The conclusion that Faraday finally reached was that lines of force must cut through a conductor in order to induce a current in it. Lines of force will cut through a conductor when the conductor is moving through a field or when a magnetic field is expanding or collapsing through a conductor. The magnitude of the current is proportional to the rate at which the lines of force are cut. The direction of the induced current is determined by the direction in which the lines cut the conductor.

A basic law from which the direction of the induced current can be determined was postulated by Lenz. It states that the magnetic field produced by the *induced* current is such as to oppose the action that induces the current. The application of this law is illustrated in Figure 2-30. A magnet is brought toward the conductor. We know that a current is induced in the wire, the question is what is the direction of the current? From Lenz's law we conclude the magnetic field produced by the current is in such a direction as to oppose the motion of the magnet toward the conductor. The direction of the magnetic field around the conductor must therefore be as shown in the figure. The direction of the conventional current flow can be determined using the right-hand rule.

These rules governing electromagnetic interactions were based entirely on experimental observations. At that time there was not as yet a comprehensive

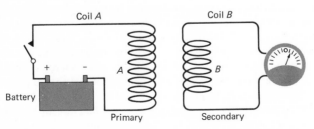

Figure 2-29 Induction. Closing or opening the switch in the primary produces a changing magnetic field which results in an induced current in the secondary.

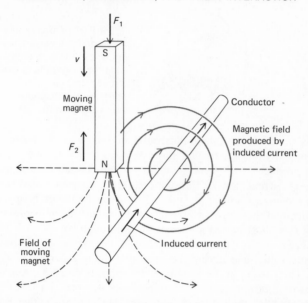

Figure 2-30 Lenz's law. The direction of the induce current is such as to oppose the change that induces the current. The magnet is moving toward the wire. Therefore the magnetic field produced by the induced current is in the direction opposing the field of the moving magnet. The induced current must flow in the direction shown.

theory which connected electricity and magnetism. A full understanding of electro magnetic phenomena was not obtained until 30 years after Faraday's pioneerin, work. The complete theory was formulated by James Clark Maxwell, whose wor we will discuss in a later section.

Most of the electrical devices used in our technology are based on the discoverie of Oersted and Faraday. Two devices, the electric generator and the transformer both invented by Faraday, are specially important in communication technology The generator produces an electric current and the transformer is used to alter th size of an electric current and to couple currents into electric circuits.

ELECTRIC GENERATOR

A simple schematic diagram of a generator is shown in Figure 2-31. A coil is mounte between the poles of a magnet. As the coil is rotated in the field it cuts the magneti lines of force and a current is induced in it. The current changes in magnitud throughout the period of rotation. The largest number of lines are cut per unit tim when the motion of the coil is in the plane parallel to the lines of force; therefor the current is at a maximum during this part of the cycle. When the coil is per pendicular to the lines of force, the current is zero since at this point in the cycl lines of force are not cut. The current is conducted from the coil by two brushe connected to slip rings. With the arrangement shown in Figure 2-31, the conventiona current in the coil flows in the clockwise direction, but if we trace the direction o the current through the brushes, we note that during each rotation of the coil the out put current changes direction. The time dependence of the output current is show in Figure 2-32. This type of time-varying current is called *alternating current*, and the specific pattern shown is called a *sinusoidal function of time*, or simply a *sine wave*. This is the type of current in common household use. In contrast to the time varying current, the current from a battery is unchanged in time and is called a *direct current*. An alternating current can be changed into a direct current by various means.

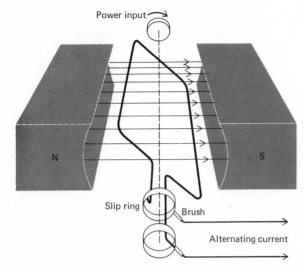

Figure 2-31 Generator. A coil of wire is shown simply as a loop rotated in a magnetic field. The induced current is conducted from the coil by brushes in contact with slip rings attached to the coil.

The current produced by the generator can be used to do work. It is therefore clear that energy must be supplied to turn the coil in the magnetic field. This is also evident from Lenz's law. The magnetic field produced by the induced current is in a direction such as to oppose the turning of the coil. The greater the current drawn from the coil, the greater the force necessary to turn the coil. The generator is a device that converts mechanical energy into electric energy.

A practical generator is much more complex than the simple model we have shown. In order to increase the effectiveness of the generator, a number of coils are rotated simultaneously in the magnetic field. The currents produced in each coil are then added together. In most generators the magnetic field is produced by an electromagnet rather than by a permanent magnet.

We have assumed so far that in all cases of a conductor moving in a magnetic field, the two ends of the conductor were connected so that a current could flow. However, if the two ends of the conductor are not connected, a current cannot flow. In that case, charges move as far as they can, which is to the end of the conductor. In this way a charge difference is produced between the two ends of the conductor. As a result of this charge difference, an electric field is produced between the two ends of the conductor. The measure of this charge build-up is the *voltage* between the two terminals. The two terminals of a generator coil are seldom connected directly to each other, but are usually connected through some device such as a light bulb or an electric motor, which in electrical engineering terminology is

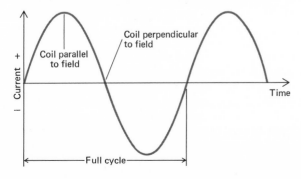

Figure 2-32 Alternating sinusoidal current obtained from generator.

called a *load*. This electric load presents some opposition to the flow of charges. Although there is a flow of current through the load, charges are still piled up at the terminals and a voltage is set up across the load. This subject is discussed more quantitatively when we deal with circuit theory.

TRANSFORMER

The principle of the transformer has already been discussed in connection with Figure 2-26. A changing current in coil *A*, which is called the *primary coil*, produces a changing magnetic field which induces a current in the *secondary coil* (coil *B*). If an alternating current flows in the primary coil, the current in the secondary coil will also alternate. Depending upon the ratio of the number of turns in the primary and secondary coils, the current in the secondary coil can be made larger or smaller than the current flowing in the primary coil. However, the voltage across the secondary coil is also changed in such a way that the product of the voltage and current in the primary is the same as in the secondary coil. As we shall see, this rule is a consequence of conservation of energy. For example, if the current in the secondary coil is larger than the current in the primary coil by a factor of two, then the voltage across the secondary coil will be half the voltage across the primary coil. Transformers are used to change high voltages into low voltages and vice versa.

3 Electromagnetic Telegraph

Following the discoveries outlined in Chapter 2, progress in communication was very rapid. The invention of the galvanometer came to the attention of Baron Schilling (1786-1837) who was in the Russian diplomatic service. He was quick to see the importance of the device for electric telegraphy — a field in which he had been interested for a long time. From 1822 until his death in 1837 he experimented with electric telegraphs and designed a number of telegraph systems. It is not clear to what extent Schilling's telegraph was put to practical use, but his work had a direct influence on the practical telegraph system developed by Cooke and Wheatstone a few years later.

One of the first electromagnetic telegraphs operated over a significant distance was that of Gauss and Weber in 1834. Karl Friedric Gauss (1777-1855) was born in Brunswick, Germany, the son of a poor peasant family. Even as a young boy he showed a great talent for mathematics. He came to the attention of the Duke of Brunswick who financed his university studies. Gauss became one of the world's greatest mathematicians. He made fundamental and original contributions to mathematics, theoretical astronomy, electromagnetic theory, and actuarial science. At the age of 30 he was appointed Director of the Observatory at Göttingen. Weber was a colleague of Gauss at Göttingen, where they started their joint telegraph project. They strung insulated copper wire over the rooftops of Göttingen, connecting the astronomical observatory to the magnetic observatory over a mile away. The signaling current was obtained from voltaic batteries and on occasions from a small electromagnetic generator. The receiver consisted of a magnet, free to swing inside the coil carrying the signal current. The direction in which the magnet swung depended on the direction of the current in the coil. The telegraph code used by Gauss and Weber is shown in Table 3-1.

There were many others who made significant contributions to the technique of electromagnetic telegraphy and built working telegraphs; but during its early stages the telegraph, as most other inventions, was a cumbersome gadget with many unsolved technical problems. In order to obtain financing and adoption of such a

Table 3-1.   The Telegraph Code of Gauss and Weber

	l = left				*r = right*		
A	*r*	M	*lrl*	0	*rlrl*
B	*ll*	N	*rll*	1	*rllr*
C,K	*rrr*	O	*rl*	2	*lrrl*
D	*rrl*	P	*rrrr*	3	*lrlr*
E	*l*	R	*rrrl*	4	*llrr*
F,V	*rlr*	S	*rrlr*	5	*lllr*
G	*lrr*	T	*rlrr*	6	*llrl*
H	*lll*	U	*lr*	7	*lrll*
I,J	*rr*	W	*lrrr*	8	*rlll*
L	*llr*	Z	*rrll*	9	*llll*

SOURCE: *Data Transmission,* W. R. Bennett and J. R. Davey. New York: McGraw-Hill Book Company, 1965. Used with permission of McGraw-Hill Book Company.

system, potential investors and backers had to be convinced of the value of the invention. The commercial adoption of the electromagnetic telegraph was pioneered by Cooke and Wheatstone in England and Morse in the United States. Although entrepreneurs like Cooke and Morse are indispensible to the development of technology, it must be remembered that the devices they champion are almost always based on a large body of scientific and technological research to which many people contributed.

William Fothergill Cooke (1806-1879), the son of a doctor, joined the East Indian army at the age of 20 but had to leave it because of ill health. He then went to study at Heidelberg in Germany. In 1836 he saw a demonstration by Professor Moncke of an electromagnetic telegraph that was a copy of Schilling's telegraph. Cooke became very interested in telegraphy and almost immediately started to construct a telegraph himself. In April 1836 he returned to England and managed to get a railway company interested in his work. Cooke himself did not know very much about the new science of electromagnetism and therefore he got in touch with Charles Wheatstone (1802-1875) who was Professor of Natural Philosophy at Kings College in London. Wheatstone at this time was doing experiments in electromagnetic phenomena and was himself interested in telegraphy. In 1837 they developed a five-needle telegraph which was patented the same year (Figure 3-1). Each magnetic needle had its own deflecting coil which caused the needle to deflect when a current passed through it. Letters were transmitted by simultaneously pointing two needles toward the given letters. Numbers were indicated by a single needle.

In July 1837 Cooke and Wheatstone demonstrated this telegraph system to the directors of the London-Birmingham Railway. Messages were transmitted between Euston and Campden Town, a distance of about 1 mile; the directors of this railway were not too enthusiastic. The following year, however, Cooke and Wheatstone installed a telegraph line between Paddington and West Drayton, a distance of about 13 miles, for the Great Western Railway Company. The operation was successful and within the next few years a number of other lines were constructed. After a rather slow start the laying of new lines increased very rapidly.

Until this time there was not much public interest in telegraphy, but an incident occurred in 1845 which drew a great deal of publicity. A woman was murdered in Slough and the murderer was seen boarding a train for Paddington. His description was telegraphed to Paddington, where he was arrested on his arrival; he was later hanged for his crime. There was now a great deal of excitement over the telegraph

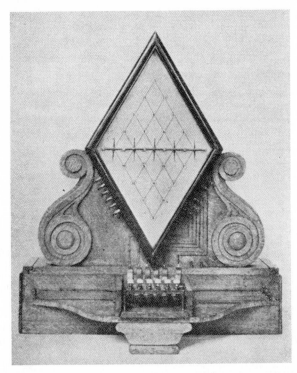

Figure 3-1 The Cooke and Wheatstone five needle telegraph.

and many public demonstrations were held (Figure 3-2). To cope with the increased demands for new telegraph lines Cooke and Wheatstone formed the Electric Telegraph Company in 1846. The company prospered and by 1852 had laid about 4000 miles of telegraph lines in England.

The telegraph of Cooke and Wheatstone was a clear extension of previous magnetic-needle deflection systems. On the other hand, the Morse telegraph system in the United States was a significant departure from the earlier schemes and eventually gained world-wide acceptance. A schematic diagram of the Morse telegraph is shown in Figure 3-3. At the sending end, key A is depressed which allows a current to flow through coil B at the receiving end. The magnetic field produced by this current attracts the pivoted stylus holder and causes the stylus to write on the moving paper strip. By interrupting the connection with key A, the "dot" and "dash" signals are transmitted and recorded. Some of the Morse code symbols are shown in Figure 3-3.

Samuel Finley Breese Morse (1791-1872) graduated from Yale University in 1810 at the age of 19. His interest in science was kindled by Professor Benjamin Silliman, but his main goal was to become an artist. After graduating he went to Europe to study painting. He returned to America, but he never became successful as a painter. He again visited Europe and apparently the idea of his telegraph system occurred to him on his return trip in 1832. In 1835 Morse became Professor of Literature of the Arts of Design at the University of the City of New York. It was during this time that he did most of his work on the telegraph. Morse was scientifically and technically inexperienced and his first telegraph was a very crude device. He probably would not have gotten very far with his telegraph system had he not obtained some very able assistance at this time. In the early parts of

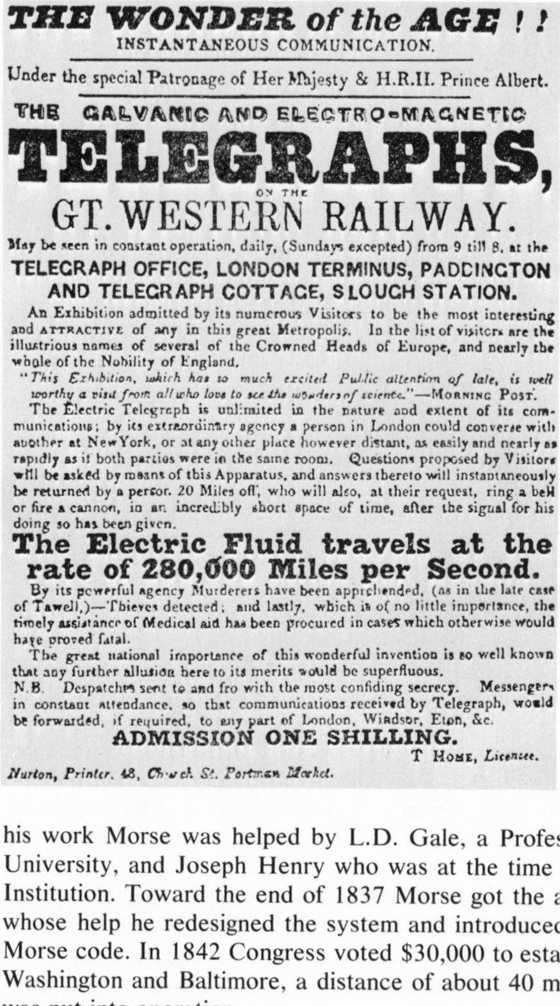

THE WONDER of the AGE !!
INSTANTANEOUS COMMUNICATION.

Under the special Patronage of Her Majesty & H.R.H. Prince Albert.

THE GALVANIC AND ELECTRO-MAGNETIC
TELEGRAPHS,
ON THE
GT. WESTERN RAILWAY.

May be seen in constant operation, daily, (Sundays excepted) from 9 till 8, at the

TELEGRAPH OFFICE, LONDON TERMINUS, PADDINGTON AND TELEGRAPH COTTAGE, SLOUGH STATION.

An Exhibition admitted by its numerous Visitors to be the most interesting and ATTRACTIVE of any in this great Metropolis. In the list of visitors are the illustrious names of several of the Crowned Heads of Europe, and nearly the whole of the Nobility of England.

"*This Exhibition, which has so much excited Public attention of late, is well worthy a visit from all who love to see the wonders of science.*"—MORNING POST.

The Electric Telegraph is unlimited in the nature and extent of its communications; by its extraordinary agency a person in London could converse with another at New York, or at any other place however distant, as easily and nearly as rapidly as if both parties were in the same room. Questions proposed by Visitors will be asked by means of this Apparatus, and answers thereto will instantaneously be returned by a person, who will also, at their request, ring a bell or fire a cannon, in an incredibly short space of time, after the signal for his doing so has been given.

The Electric Fluid travels at the rate of 280,000 Miles per Second.

By its powerful agency Murderers have been apprehended, (as in the late case of Tawell,)—Thieves detected; and lastly, which is of no little importance, the timely assistance of Medical aid has been procured in cases which otherwise would have proved fatal.

The great national importance of this wonderful invention is so well known that any further allusion here to its merits would be superfluous.

N.B.   Despatches sent to and fro with the most confiding secrecy. Messengers in constant attendance, so that communications received by Telegraph, would be forwarded, if required, to any part of London, Windsor, Eton, &c.

ADMISSION ONE SHILLING.
T HOME, *Licensee.*

Norton, Printer. 48, Church St. Portman Market.

Figure 3-2 An announcement of a public exhibition of telegraphy (1845). (Courtesy of the New York Public Library.)

his work Morse was helped by L.D. Gale, a Professor of Chemistry at the City University, and Joseph Henry who was at the time Secretary of the Smithsonian Institution. Toward the end of 1837 Morse got the assistance of Alfred Vail with whose help he redesigned the system and introduced the now famous "dot-dash" Morse code. In 1842 Congress voted $30,000 to establish a telegraph link between Washington and Baltimore, a distance of about 40 miles. On May 24, 1844, a line was put into operation.

In the United States just as in England progress in building telegraph links was very rapid. Many companies were started and the competition was fierce and chaotic. In 1856 there was a great amalgamation of companies, and Western Union

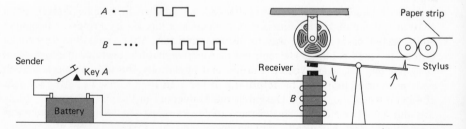

Figure 3-3 Morse telegraph. When key *A* is depressed, the electromagnet *B* at the receiver is energized and attracts the pivoted stylus arm. The stylus then draws a line.

was formed. This period of great growth was a fascinating one. It was filled with patent battles, financial scandals, and different groups fighting for rights of way for their telegraph lines.

TRANSOCEANIC TELEGRAPHY

With telegraph lines connecting large sections of England, continental Europe, and the United States it was natural to start thinking about transoceanic connections. The installation of these cables required incredible perseverance. Cable technology was very poor and there was virtually no knowledge about the configuration of the ocean bottom. Even if connections were successful, it was known that the signals would be extraordinarily weak, maybe not even detectable. Yet people tried. The first attempt at laying a submarine telegraph cable was a modest one. On August 28, 1850, a cable was laid between England and France under the direction of two brothers, John and Jacob Brett. The cable was not strong enough and broke. A new, stronger cable was manufactured and in September 1851 another attempt was made, this time successfully. Cables were then laid (often unsuccessfully) in the North Sea and the Mediterranean, but the biggest challenge still remained, a cable connecting Europe and the United States.

Most of the credit for the transatlantic cable is due to Cyrus W. Field (1819-1892), an American businessman. In 1856 he and two Englishmen founded the Atlantic Telegraph Company. His two partners were John Brett (who was involved in the England to France link) and Charles Tilson Bright, an engineer with the British Electric Telegraph Company. The English Government agreed to provide £14,000 in annual subsidy to the company and it was assumed that the United States Government would provide a similar sum. In Congress, however, there was a surprising amount of opposition to the venture. The whole idea of establishing such close contacts with England was questioned, but the project was finally given congressional approval in March 1857.

The manufacture of the cable was an unprecedented task. It was made of a stranded inner copper conductor insulated with guttapercha and tarred yarn. The whole cable was then surrounded by stranded steel cables for outer protection. A total of over 17,000 miles of copper wire was used and the tarred yarn alone weighed 240 tons. The first laying of the cable was attempted in August 1857. The cables were carried by the United States frigate, *Niagara,* and the British battleship, *Agamemnon.* After 380 miles of cable were laid, the weather became bad and the cable broke. Another attempt was started on June 10, 1858. Again the sea became stormy; there were a number of serious injuries aboard the *Agamemnon.* An electrical break was discovered in the cable that had already been laid and the attempt again was a failure. The ships reassembled and began to lay the cable once more. After further difficulties, including an attack on the cable by a whale, the link was completed, 2000 miles of unbroken electric cable now connected Valencia, an island, to Trinity Bay in Newfoundland. There was great elation over the project. A 90-word message from Queen Victoria to President Buchanan was transmitted. Unfortunately after a few weeks the insulation on the cable became damaged and the cable stopped transmitting. Naturally all who were concerned with the project were greatly disappointed. There were many who claimed that, in fact, communication was never established and that the whole thing was a fraud. Work was discontinued for a few years, but by 1864 enough money was raised to start another attempt. In the summer of 1865 the line was nearly completed when the cabl slipped and was lost again. At last on July 27, 1866, a successful and permanen link between the continents was made. This time the project went well and the cabl that had slipped during the previous year was retrieved and also put into operation

Following the successful laying of the transatlantic cable, other government and private investors gained confidence in this type of project and within a fev decades most of the major world population centers were interconnected by tele graph cables.

MODERN TELEGRAPHY

The basic principle of telegraphy is simple. A coded current is sent through ; pair of wires from the sender to the receiver, where the message is decoded. (Tw(wires are necessary because a return path for the current must be provided.) As th use of the telegraph increased it quickly became evident that it was not practica to have separate pairs of wires connecting each user of the system. By the 1860(central telegraph offices were built in major cities. Each local telegraph statio was connected to the central office in its region by one pair of wires. The loca office sent its message to the central office, which relayed it to its destination Depending on the destination, the message may have passed through a numbe of centrals before reaching the final receiver. In this simple and obvious way ; telegraph office could communicate with all other offices in the world.

Toward the end of the last century, teletypewriters were developed that in creased the speed of transmission and decreased the number of human operator necessary to handle the system. Using a teletypewriter the operator depresses ; typewriterlike key on the keyboard. This activates a set of electromagnets calle(*relays,* which automatically switch on electric circuits and generate the electri(pulse sequence corresponding to the letter on the key. At the receiving end thes(pulses energize a set of relays which then cause the transmitted letter to be type(on paper. Transmission and reception of signals by machines require a differen coding from the previous Morse code. In the Morse code each letter is designate(by a unique sequence of short and long pulses, dots (·) and dashes (-). The numbe of units is not the same for each letter. For example, E is designated by one shor unit (·), whereas Y is designated by four units (- · - -). In the machine code th(pulses are of the same size. All letters are represented in a five-unit code; that is five pulse intervals are alloted for each letter. The letters are coded by the positio and number of pulses within this interval.

The early teletypewriters, of course, required an operator to type the letters The system was not as efficient as it could have been because the machine wa: able to transmit messages much faster than the operator could type them. A grea increase in the speed of transmission was brought about by the introduction o punched tapes. The operator types the message on a machine that perforates a pape tape with holes. A different combination of holes across the tape represents eacl letter of the alphabet. The perforated tape is then fed into a transmitter machin(in which the holes in the tape are probed by metal fingers. The fingers close electri(circuits through the perforated parts of the tape producing the pulsed code signa for transmission (Figure 3-4). At the receiving end the message is either typed b) a teletype machine or stored on tape for later display. The speedup of transmissio

Figure 3-4 Teletypewriter. (Courtesy of International Business Machines Corporation.)

occurs because a tape transmitter can transmit messages at a rate more than 20 times faster than the transmission rate of a manually operated teletypewriter. One tape transmitter can therefore process tapes from many operators.

The speed and capability of the system were further improved by the introduction of automatic switching between transmitting machines. This allowed messages to be quickly routed from one station to another. Recently computers have been introduced into the telegraph system. The message, instead of being recorded on perforated paper tape, is stored in the computer. The computer automatically groups the messages and then feeds them to the transmitter. In many installations the telegraph system has been made completely automatic.

When the telephone was first introduced into public service, it appeared for a while that telegraphy might be completely replaced by the newer device. This did not happen. The technological improvements in telegraphy, some of which we have just described, gave the telegraph two important advantages over the telephone. Messages can now be transmitted at a much higher rate by telegraph than by telephone and a permanent record of the message is produced automatically. This is especially important in cases where the message concerns a number of people. Today the telegraph is widely used in news, diplomatic, and business message transmission. In a typical application, for example, each branch office of the user is equipped with a teletypewriter through which messages can be relayed and received to and from the central office. In addition, through an automatic central office the local office can contact any other teletype installation in the world. The teletype facilities of the U.S. State Department often receive as many as 7000 diplomatic messages a day. The printed messages are automatically duplicated and distributed to an average of 100 people for evaluation. Clearly such efficiency would be hard to obtain with a telephone.

4 Telephone

With the telegraph completed inventors' interests next turned to the problem of transmitting sound. Sound is a mechanical phenomenon produced by vibrating objects. When an object such as a tuning fork or the human vocal chords is set into vibrational motion, it transfers this motion to the surrounding air molecules. These molecules follow the motion and transfer it to adjacent molecules. In this manner the disturbance propagates. When the air vibrations reach the ear, they set up vibrations in the eardrum, which then become nerve impulses and are interpreted by the brain. A medium is needed between the source and the receiver in order to propagate sound. This is demonstrated by the well-known experiment of the bell in the jar (Figure 4-1). When the bell is set in motion, its sound is clearly audible. As the air is evacuated from the jar, the sound of the bell diminishes and finally becomes inaudible.

Materials other than air also transmit sound. In fact, many materials are much better transmitters of sound than air. Sound travels with the speed characteristic of the propagating medium. In air at atmospheric pressure sound travels at a speed of 1087 feet per second. In iron it propagates much faster, at 16,820 feet per second.

Communication directly by sound is very inefficient. The propagating medium

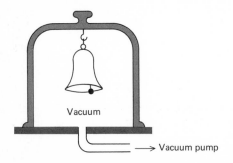

Vacuum

→ Vacuum pump

Figure 4-1 A ringing bell in an evacuated bell jar cannot be heard.

43

has to be set into vibrational motion and because the propagating materials are not perfectly elastic, the motion is quickly dissipated by friction. In the late 1600 Robert Hook experimented with the propagation of sound along a taut wire. He was able to transmit a voice over about 300 yards of wire, but beyond that the sound was inaudible. With the discoveries of Faraday in 1831 it became possible to trans late the information content of a sound wave into an electric current signal, which could then be transmitted by conducting wires to its destination. There it could be converted back into sound. Because the loss of signal strength is smaller, an electric signal can be conducted over much larger distances than sound. In order to understand how sound is converted into an electric signal we must examine in greater detail the mechanics of sound.

We have already stated that sound is created by vibrating bodies. The vibration of the sound source cause alternate compression and rarefaction in the sound conducting medium. The vibrating molecules near the sound source transfer the motion to adjacent molecules and in this way the compressions and rarefaction propagate away from the source. This train of alternate compressions and rarefac tions is called a *sound wave* (Figure 4-2). Two important characteristics of sound are the intensity, which is determined by the magnitude of compression and rarefac tion in the propagating medium, and the frequency, which is determined by how often the compressions and rarefactions take place. For convenience we represent the density variations in a sound wave by a simplified plot of pressure versus distance or time. Figure 4-2 shows the variation of molecular density or pressure in a sound wave and its graphic representation. The zero axis in the figure represents the molecular density in the absence of a sound perturbation. The sound wave causes the density and pressure to vary about this value. In the regions of com pression the molecular density is above normal and in the regions of rarefaction

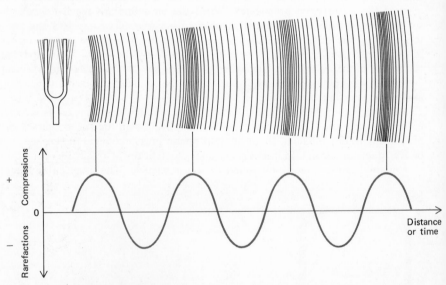

Figure 4-2 Propagation of sound waves. The vibrations of an object produce alternate com pressions and rarefactions of the medium. These compressions and rarefactions propa gate away from the source.

it is below normal. If we monitor the density variations at a fixed point in the medium, we find that the density varies with time, the air being alternately compressed and rarefied as the wave propagates through the medium.

The concepts of intensity and frequency are illustrated in Figures 4-3 and 4-4, respectively. The intensity of the sound is determined by how much the molecules are compressed in the sound wave. If the source of sound vibrates with a large amplitude, the compressions and rarefactions are large and therefore the intensity of the sound is great. The frequency of the sound is determined by how many times the sound source vibrates per second. This in turn depends on the construction of the vibrating body. When struck a large object vibrates more slowly than a small one, and the sound it produces is of lower frequency than the sound produced by the small object. The frequency of sound is referred to in music as the *pitch*. For example, the frequency of the standard A note is 440 vibrations per second. The frequency of the C below this is 264 vibrations per second.

The speed with which the wave propagates depends on the elastic properties of the medium. The more elastic the medium, the faster the displaced molecules return to their initial positions, thus resulting in a greater speed. The intensity of the sound decreases with distance from the source. This is due to the dissipation of sound energy in the medium and the inevitable spreading of the sound wave. Because the sound wave spreads as it moves away from the source, the force available to move a given volume of the propagating medium is reduced. The spreading of the sound can be reduced by focusing the sound with a horn or channeling it with a wire, but the spreading cannot be entirely eliminated.

In Figure 4-2 we showed the sound produced by a tuning fork. Here the sound wave consists of rather simple amplitude variations which are called *sine waves*. (This is the same type of a function as is produced by the electric generator.) The pressure variations in the sound of the violin or human voice are much more complex. The wave forms of different instruments playing the same note are shown in Figure 4-5. It is the detailed structure of these wave patterns that differentiates one musical instrument from another. In a subsequent section we shall discuss this point in greater detail.

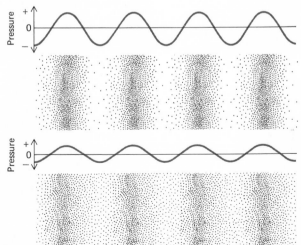

Figure 4-3 Intensity of sound waves. The intensity of sound waves depends on the degree of compression (densely spaced dots) and rarefaction of the propagating medium. The top sound wave is twice as intense as the bottom one.

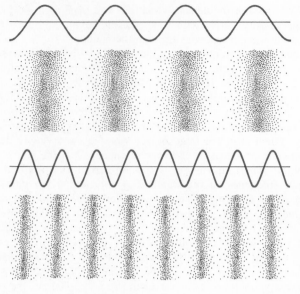

Figure 4-4 Frequency of sound waves. The frequenc of sound waves is the number of compressions and rarefactions passing a given point in one second (cycles per second). The frequency of the bottom wave is twice that of the to wave.

PRINCIPLES OF TELEPHONY

The conversion of sound into an electric signal is based on the discovery of Farada that a moving magnet induces in a coil a current that is proportional to the motio of the magnet. This is the operating principle of a microphone (Figure 4-6). membrane is attached to a magnet which can move freely inside a coil of wire. Pres

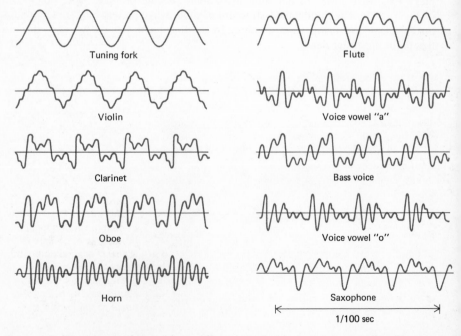

Tuning fork

Violin

Clarinet

Oboe

Horn

Flute

Voice vowel "a"

Bass voice

Voice vowel "o"

Saxophone

1/100 sec

Figure 4-5 Wave forms of sound from different musical instruments sounding the same not

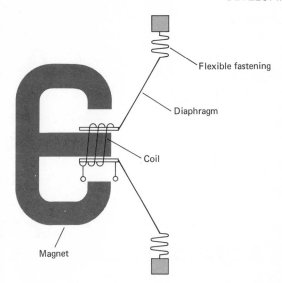

Figure 4-6   Microphone. Pressure variations due to sound produce motion of the diaphragm and the attached coil. The coil moves in a magnetic field and therefore a voltage proportional to the sound is induced in the wires. In principle the construction of a speaker is identical to that of a microphone. Here a voltage applied to the coil causes the coil and the attached diaphragm to move. The diaphragm produces motion in the surrounding air.

sure variations of the sound cause the magnet to move back and forth inside the coil. Because the motion of the magnet is proportional to the pressure variations caused by sound, the current induced in the wire is also proportional to the sound. An intense sound produces a large displacement of the magnet and therefore a large current. Similarly, a low sound produces a small current. In this way the fluctuations of pressure in the sound wave are transformed into corresponding fluctuations of current. This current is then conducted to the receiving station, where it is converted back to sound by a receiver or speaker.

The conversion of an electric signal into sound is based on Oersted's discovery that a current produces a magnetic field. The basic construction of a speaker is the same as the construction of a microphone (Figure 4-6). The signal current flows through the coil and produces a magnetic field that causes the diaphragm to move. The motion of the diaphragm follows the current variations and sets up pressure waves that are audible. Since the speaker diaphragm duplicates the motion of the microphone magnet, the sound it produces is identical to the driving sound. In an actual speaker the current-carrying wire is wound around an iron core to strengthen the magnetic field. There are many other microphone and speaker designs, some of which we shall discuss later. The system we described is the one that was patented by Bell (Figure 4-7). In Bell's telephone there is a battery in a circuit that provides continuous current through the circuit so that the current variations due to the sound occur about this mean value.

EVELOPMENT OF THE TELEPHONE

Although most people associate the telephone with Alexander Graham Bell, many others made significant contributions to the development of the device. In fact the first telephone was built in 1860 by Philipp Reis (1834–1874). This was 15 years before Bell's invention. His microphone was based on a principle somewhat different from the one we discussed, but the device was sound and his telephone worked. It is reported that at a demonstration he recited "Ach du Lieber Augustin" through his telephone. But as is the case with most inventions, the entrepreneur

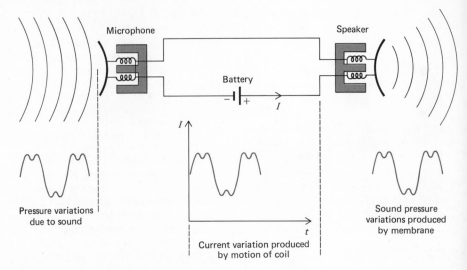

Figure 4-7 The operation of a telephone.

plays a vital role in the success of a device and Reis was not a good entrepreneu He was a German schoolteacher and not a member of the established Germa scientific community. Reis was not able to obtain financial backing and his ir vention died. Credit for the first practical telephone goes to Bell.

Alexander Graham Bell (1847-1922) was born in Edinburgh. As a young ma he taught elocution in London, but in 1870 he and his family moved to Canad; He did not know much about science or electricity, but throughout this perio he was very interested in the problem of sound. He taught deaf mutes and ex perimented with devices for the deaf. In 1872 he started a school for the correctio of stammering in Boston. During this period Bell worked on various devices fc the deaf, and this led him to the idea of the telephone in 1874. He obtained som financial backing from local businessmen and the aid of a very capable instrumer maker, Thomas A. Watson. Together Bell and Watson worked on the technic; details of the telephone. On February 14, 1876, Bell applied for a patent on h device and a month later on March 10, 1876, in Boston, Bell made his first tel(phone conversation. He spoke to Watson saying, "Mr. Watson, come here. want you." It is reported that this was not what Bell intended to say as his fir message. He was preparing to recite a passage from Shakespeare when some batter acid accidentally spilled on his clothes. It was then that he shouted into the tel(phone the now famous message.

On the same day that Bell applied for the patent, but two hours later, Elish Gray (1835-1901) applied for a caveat on a telephone system that he had d(veloped independently of Bell. Gray was better qualified in electricity than Bel Prior to his work on the telephone he had already made significant contributio to telegraph technology. His microphone was quite different from Bell's origin; idea. In Gray's microphone the sound wave impinging on a diaphragm varied th electric resistance of the circuit and in this way caused current variations in syr pathy with the sound signal. The caveat for which Gray applied was not a pater application. It was a notification to the patent office that the inventor was diligentl pursuing an idea which was not fully developed. At the time of his patent aj plication Bell had not transmitted the human voice and when he did communicate with voice, he actually used Gray's microphone and not his own. In subsequent patent litigations it was claimed that Bell had seen Gray's caveat and then copied Gray's invention, but Bell claimed that this was not so and that he himself had long ago thought of this idea. Bell was granted the telephone patent.

In the following few years many improvements were made in telephone technology by Hughes, Blake, Dolbear, Berliner, Pupin, and many others. In 1877 Edison patented a greatly improved variable resistance carbon microphone which in a modified form is still used today.

In 1877 Bell established the Bell Telephone Company. A year later the first telephone exchange was opened in New Haven. The telephone developed rapidly in many countries. In Britain the first commercial telephone lines were established in 1878 using mostly equipment imported from the United States. From the very beginning in Britain the telephone development was under government control by the general post office. Through investments of the American Telephone and Telegraph Company, which was set up by the Bell interests in 1885, telephone communications were built in France, Belgium, Spain, and Germany.

There was still a limitation to telephone communications. It was restricted to relatively short interurban distances. The electric currents generated by the microphone became too weak to drive the telephone speakers at the receiving end after traveling a few hundred miles. This problem was solved by the development of electronic amplification. On January 25, 1915, using newly developed amplifiers, Alexander Graham Bell again held a telephone conversation with Thomas Watson, but this time across the continent from New York to San Francisco.

MODERN TELEPHONY

We have described the successful laying of a transatlantic telegraph cable in the 1860s. It took nearly another 100 years for a transatlantic telephone cable to be laid. The reason for this was the lack of suitable amplifiers. As we have mentioned a telephone signal propagating along a wire has to be amplified at 100-mile intervals if it is to remain strong enough to operate a receiver. Although amplifiers were already available in the early 1900s, they required much more development before they were adequate for underwater operation. Since a failure of an amplifier in an underwater cable requires the pulling up of at least a section of the cable, repairs are very expensive. The underwater amplifier must therefore be very reliable. Currently used underwater amplifiers are expected to operate at least 20 years without servicing. In addition, the amplifier has to be watertight and able to withstand the great water pressure under the sea.

The first underwater telephone cable with amplifiers was laid in 1943 in the Irish Sea between Anglesey, Wales, and the Isle of Man. The cable was laid in relatively shallow water, which made servicing easier and construction requirements less stringent. Following this venture, a number of other short underwater telephone cables were laid and finally in 1953 a telephone cable connection was completed between North America and Europe. Today there are hundreds of underwater telephone cables reliably interconnecting all parts of the world.

A central exchange is required for telephones for the reason we discussed in connection with the telegraph. The first exchanges were manually operated. The

person making a call picked up the receiver which operated a pair of contac
causing a light to flash in the exchange. The operator then spoke to the send
and activated the bell of the telephone at the receiving end. When that telepho
was answered, the operator plugged in the sender's wires to those of the receiv
and the call was completed. This system was adequate for as long as the numb
of subscribers was relatively small. But as the number of telephone users gre
and electronic amplification increased the range of telephone contact, the manu
switchboards were not able to cope with the load. By about 1912 automatic switc
ing techniques were developed and introduced into some of the busy central e
changes. This innovation increased enormously the capacity of the telepho
system. The first automatic switching systems were electromechanical. The diali
of the numbers opened and closed electric contacts which generated pulses th
were transmitted to the central switching station. Here the signals activated t
appropriate relays to complete the connection to the telephone with the dial
number. Although in principle the system is simple, the switching is, in fact, ve
complex because of the large number of subscribers tied into the network. T
system must provide for alternate routing of the call when the most direct co
nection is used by another subscriber. It must also first signal the subscriber th
he can start dialing (the dial tone) and then indicate if the connection is complet
or not (the ring or busy signal). The system has been further complicated by t
introduction of direct long-distance dialing and other services provided by t
telephone company.

The electromechanical switching systems are now being replaced by an entire
electronic system. The relays are being replaced by solid state switches whi
operate without mechanical contact. The new system is faster and less noi
than the older one. Since response of the solid state switches is faster, the co
ventional dial is replaced by push buttons that can generate the number code mo
rapidly than the dial. Computers which have been recently introduced into t
system have reduced the cost of billing and have further increased the efficien
of the telephone system.

5 Wireless Communication

BASIC PRINCIPLES

The theoretical foundation of the radio was put forth in 1864 by James Clark Maxwell (1831-1879), a great theoretical physicist of the nineteenth century. He was born in Edinburgh, the son of a lawyer whose hobby was scientific gadgets. A brilliant student, especially in mathematics and science, he attended the University of Edinburgh and later Cambridge. While at Cambridge he experimented with sleep and exercise routines. For a prolonged period of time he slept from 5 to 9:30 P.M., then read and studied until 2 o'clock in the morning. Between 2 and 2:30 he exercised and then slept again from 2:30 until 7:30 A.M. He found this routine very effective. Soon after his arrival at Cambridge, Maxwell became a teacher at Trinity College but reportedly was a very poor lecturer. However, his research interests were amazingly broad. He experimented with cats trying to find out at what age they developed the ability to land on their feet. In his work with colors he showed that the combination of red, green, and blue could produce any color. University of Cambridge offered a prize for the explanation of the rings around the planet Saturn; the question was whether the rings were made up of solid, liquid, or dust. Maxwell won the prize for a brilliant mathematical paper showing that only disconnected particles would form the stable ring observed around Saturn. In 1864 he published the first of a series of articles on electromagnetic theory which formed the basis of radio communication.

Maxwell's work in electromagnetism was purely theoretical. He examined all the electromagnetic phenomena discovered by Oersted, Faraday, and others and formulated mathematical equations to explain them all in a unified way. His equations summarized all the known electromagnetic laws and interactions and in addition predicted some new important effects.

It is interesting to see how Maxwell obtained these equations. His was an age of mechanics and it is therefore not surprising that he actually postulated a mechanical model which in an analogous manner displayed the electromagnetic phenomena. He filled space with imaginary gears, the motions of which represented flows of currents and magnetic fields. With this model in mind, he described its

51

behavior with mathematical equations. Maxwell, of course, knew that this mechanical model had no physical reality and that it was only a scaffold for the theory. Maxwell's equations are among the most important theoretical developments in modern physics. Although physics has undergone great changes during the last century, Maxwell's theory remains valid without alterations.

In Maxwell's theory the electromagnetic interactions are viewed entirely in terms of fields. Let us see how this is done for the two most important interactions found by Oersted and Faraday. Faraday found that a changing magnetic field induces a current in a conductor. But an electric current is electric charges put into motion, which requires that a force be applied to the charges. Therefore we conclude that a changing magnetic field produces a force on electric charges. As we mentioned before, force is exerted on electric charges by electric fields. Through this chain of reasoning we conclude, finally, that a changing magnetic field creates an electric field.

Maxwell introduced a symmetry into electromagnetic theory by postulating that a changing electric field produces a magnetic field. Oersted's discovery follows very simply from this postulate. Oersted found that a current produces a magnetic field. But current is simply charges in motion and these charges are surrounded by electric fields that are also in motion and therefore changing. In this view Oersted's magnetic field is created by the changing electric field of moving charges.

In summary, a changing magnetic field creates an electric field and a changing electric field creates a magnetic field. These two statements clearly point out the remarkable symmetry between electric and magnetic phenomena. Yet the symmetry is not complete. We recall that it was this lack of symmetry that disappointed Faraday. He was able to produce a steady magnetic field with a current, but with magnets he could produce only transient and alternating electric fields. This is due to the fact that positive and negative charges exist as separate entities, but magnetic poles do not. If separate magnetic poles existed and we could make a current of them, then this magnetic current would be surrounded by a steady electric field and the symmetry would be complete.

We may next ask what is the direction of these fields produced by the changing fields? The answer can be found from Oersted's and Faraday's experiments, but it has been most concisely stated by Maxwell: The electric field is created perpendicular to the direction of the *change* in the magnetic field, and the magnetic field is created perpendicular to the direction of the *change* in the electric field. There is a new concept here: Not only does a field have direction, but the change in that field also has a direction. As shown in Figure 5-1 the change in the field need not be in the same direction as the field itself. The interrelation between electric and magnetic fields is shown in Figure 5-2.

Based on the equations for the electric and magnetic fields, Maxwell made a most important prediction. He showed that it should be possible to produce an electromagnetic field that propagates away from the source. Such a propagating field is called an *electromagnetic wave*. The process is illustrated in Figure 5-3.

$\longrightarrow$	$\longleftarrow$	$\longrightarrow$
Magnetic field at time t_1	Change in magnetic field during time interval $t_2 - t_1$	Magnetic field at a later time t_2

Figure 5-1 The change in the magnetic field need not be in the same direction as the magnetic field.

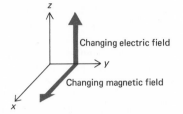

Figure 5-2 A changing electric field produces a changing magnetic field and vice versa.

Let us consider a charge at the end of a rod. When the charge is stationary, an electric field emanates from the charge; there is, however, no magnetic field around the charge. But when the charge is put in motion, a current is produced by the moving charge and therefore the charge is surrounded by a magnetic field. If the charge has an oscillatory motion, then within each cycle the current and the associated magnetic field reverse direction. The changing magnetic field produces an electric field that is perpendicular to the magnetic field. Since the source of this electric field is continuously changing, the electric field is also changing, which in turn produces another magnetic field. In this way a propagating electromagnetic wave is produced in which the changing electric field produces the changing magnetic field, which in turn produces another changing electric field, and so on. The electromagnetic wave is in this way self-reproducing and detached from the source. The frequency of the field, or the number of times per second that the electric and magnetic fields undergo a complete change, is determined by the oscillating frequency of the source change. Maxwell showed that in empty space the electromagnetic waves propagate at a constant speed independent of their frequency. On the other hand, in a material medium the speed of propagation depends on the frequency and is generally less than the speed in empty space.

At the time that Maxwell developed these ideas many of the properties of light were already well known. It was known, for example, that light travels more slowly in materials such as glass than it does in empty space. The various interference properties of light such as the light-dark alternating regions produced when a beam of light passes through a narrow slit were also known. Maxwell pointed out

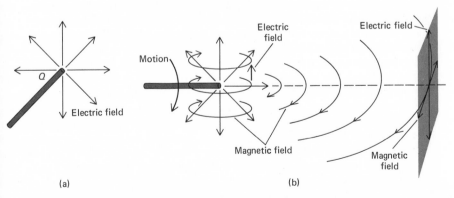

(a)   (b)

Figure 5-3   (a) A stationary electric charge is surrounded by an electric field but no magnetic field. (b) If the charge is set into motion, a current results which produces a magnetic field. The back and forth motion of the charge results in a changing current and changing magnetic field. The changing magnetic field produces a changing electric field, which in turn produces a changing magnetic field, and so on. A small part of the electric and magnetic field detaches from the source and propagates into space.

that his postulated electromagnetic waves exhibit all the properties of light ar he suggested that light is also an electromagnetic wave. The characteristic spee at which electromagnetic waves propagate is therefore just the speed of ligh which was by then known to be about 186,000 miles per second.

When Maxwell first published his equations governing electromagnetic phe nomena, there were very few people who understood them. His work was elega and concise but way ahead of his time. The theory was fully explained and verifie by Heinrich Hertz (1857-1894) in 1887, more than 20 years after its publicatio At that time the key point in Maxwell's equations was the prediction that lig and electromagnetic waves of the type generated by oscillating currents were th same phenomenon. It is this that Hertz set out to prove. Later in his career in popular lecture Hertz told his audience, "I am here to support the assertion th light of every kind is itself an electrical phenomenon. The light of the sun, th light of a candle, the light of a glowworm."

What then is the difference between light and radio waves? We certainly kno that there are differences in their behavior. For example, light is blocked by objec through which radio waves penetrate without any difficulty. We can see objec that are illuminated by light, but our eyes are not sensitive to radio waves. W might also include in this discussion the difference in properties of x rays, cosm rays, infrared rays, and microwaves — they all are electromagnetic radiation. Th different behavior of these radiations is due entirely to the difference in their fre quencies.

The whole spectrum of electromagnetic radiation is shown in Table 5-1. Electr magnetic waves, and all other waves, are characterized by three parameter frequency, wavelength, and speed of propagation. The frequency is simply th number of times the field changes direction through a full cycle in 1 second. W will use the symbol f for frequency. The period of a wave, which we designat by the letter T, is the time required to complete one full cycle of oscillation. Sinc f cycles are completed in 1 second, one cycle is completed in $1/f$ seconds. In oth words, $T = 1/f$. The wavelength is a length the wave travels during one perio Traditionally the Greek letter λ (lambda) is used to designate wavelength and the speed of propagation.

The parameters c, λ, and f are related. We know that the distance D covere in the T by any entity (object or wave) moving at a velocity v is given by

$$D = Tv$$

Therefore the distance λ traversed in a time T by a wave moving with velocity is

$$\lambda = Tc \qquad (5\text{-}$$

But for T we may substitute $1/f$, which gives us the relationship of the parameter c, λ, and f, namely

$$\lambda f = c \qquad (5\text{-}2$$

Electromagnetic waves can be represented graphically as we have done wit sound. In Figure 5-4a we show the wave as a function of distance viewed at a fixe time. The vertical axis indicates either the electric or magnetic field strengt The figure also shows an equivalent and possibly more direct definition of wave length than we gave above. Here the wavelength is shown as the distance be tween two closest equal amplitude points on the wave train.

Table 5-1   The Electromagnetic Spectrum

f Frequency (sec^{-1})	λ Wavelength (cm)		
10^{32}	3×10^{-22}		
10^{30}	3×10^{-20}	Cosmic	
		ray	
10^{28}	3×10^{-18}	photons	
10^{26}	3×10^{-16}		
			← Limit of accelerator energy
10^{24}	3×10^{-14}		
			← Wavelength = size of elementary particle
10^{22}	3×10^{-12}		← Limit of nuclear gamma rays
10^{20}	3×10^{-10}	Gamma rays	← Limit of atomic rays
10^{18}	3×10^{-8}	x rays	
10^{16}	3×10^{-6}	Ultraviolet	Visible light
10^{14}	3×10^{-4}	Infrared	
10^{12}	3×10^{-2}	Microwaves	
10^{10}	3	Radar	
		UHF	
10^{8}	300	VHF, FM	
		Shortwave	
10^{6}	3×10^{4}	AM radio	
10^{4}	3×10^{6}	Longwave	
		radio	
10^{2}	3×10^{8}		← Wavelength = radius of earth

At any point in space the amplitude of the field changes with time. This is shown in Figure 5-4b. The patterns in the space and time plots are, of course, identical. The time interval between points of equal amplitude is the period, and the frequency is the number of periods in a second.

The wave pattern shown in Figure 5-4 is the sinusoidal function we have encountered before. It is a special wave shape generated by regularly oscillating charges.

We can see from Table 5-1 the enormous range of electromagnetic wavelengths found in nature. The wavelengths extend from the unimaginably small 10^{-22} centimeters to many kilometers. We know from Maxwell that all of these wavelengths are produced by the same phenomena, namely accelerating or oscillating electric charges. Let us look at how the same process can produce such a wide variety of wavelengths and frequencies. Consider a charge q which is to be put into oscillatory motion through a distance d (Figure 5-5). The length of time for the completion of the full oscillation cycle from point A to B and back to A is $2d/v$. Here v is the average velocity with which the charge is moving. This time

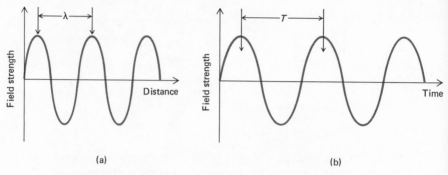

(a)   (b)

Figure 5-4   (a) Field strength as a function of distance. The wavelength (λ) is the distance be
tween adjacent peaks. (b) Field strength as a function of time. The period T is the tim
interval between successive peaks.

interval is, of course, the period of oscillation. Substituting this into Equatio
5-1, we get

$$\lambda = \frac{c}{v} 2d$$

From Einstein's theory of relativity we know that a charge cannot move any faste
than the speed of light; therefore c/v is always larger than one. Using this fact i
the equation we just derived, we conclude that the wavelength of the electro
magnetic radiation is always longer than the distance over which the source charg
oscillates. Antennae with dimensions on the order of centimeters produce radic
waves, but light waves can be produced only if the charges are moved throug
distances shorter than the wavelength of light, which is about 5×10^{-5} centi
meters. It is not possible to manufacture an antenna this small; however, the motio
of electrons within the atoms is on this order. Electromagnetic radiation in the x-ra
and light wavelength regions is emitted by electrons within atoms. The still shorte
wavelength gamma rays are produced by the motion of charges within the nucleus.

After this detour, let us return to Hertz's experiment. His equipment is show
schematically in Figure 5-6. When switch A is opened, the current through th
primary coil is interrupted and as a result a large electric field is induced in th
secondary coil. This large field causes charges to jump across the gap and oscillat
back and forth across it (see section on induction, page 72). As the current flow
across the gap it heats the air in the gap causing it to glow. In this way the pat
of the current is visible as a spark. The pulse of oscillating current generates
pulse of electromagnetic field, a portion of which propagates away from the spar
gap. In Hertz's experiments sizes of the components were such that the wavelengt
of the radiation was about 30 centimeters.

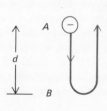

Figure 5-5   The charge
placed into an oscillatory
motion between points A
and B. The wavelength of
the radiation emitted by the
charge is shown to be
longer than the path length
of oscillation.

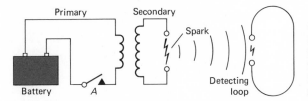

Figure 5-6 Hertz's experiment. The oscillating current in the spark produces electromagnetic radiation which in turn produces a small spark in the detecting loop.

The detection of the electromagnetic wave is simple in principle. If a conductor is placed in the path of the propagating field, the electrons in the receiving conductor are forced into motion, producing a current that can be detected. For detection Hertz actually used a loop of wire with a small gap in it. He detected the current caused by the field by observing a spark in the gap.

With this simple apparatus Hertz demonstrated that electromagnetic radiation produced by the motion of charges behaves as light. By putting a conducting surface in the path of the electromagnetic wave he showed that the wave is reflected by the surface as light is reflected by a mirror. He also demonstrated that electromagnetic radiation can be focused and that it propagates at the speed of light. Any differences between the properties of light and the electromagnetic waves generated in Hertz's apparatus could be explained as due to the difference in their wavelengths. For example, it was well known that because of diffraction, light is bent by a glass prism. Hertz was able to show that his radiation is also bent by a prism, but because of the enormously larger wavelength, he had to use a prism made of about half a ton of pitch. Light is obstructed by material such as wood or paper, but Hertz found that again because of the long wavelength, these materials were transparent to the radiation in his experiment. The importance of Hertz's experiment was recognized immediately, and soon his experiments were reproduced and extended in laboratories all over the world.

Having made these brilliant contributions to electromagnetism, Hertz abandoned this line of work and turned to experiments in atomic physics, but unfortunately his work in this field was cut short. He died in 1894 at the age of 37.

BEGINNING OF WIRELESS COMMUNICATION

Although Hertz himself did not think that his findings were of any practical use and argued against the possibility of using electromagnetic waves for communication, there were many people who saw immediately the potential in Hertz's demonstration. Electromagnetic waves are, in fact, an excellent means of communication. Wire connections are not needed between sender and receiver and the speed of communication is nearly instantaneous.

The principle of using electromagnetic waves for communications is actually contained in Hertz's experiment. It is summarized in Figure 5-7. At the transmitting end, charges are set into oscillatory motion in a conductor called a *transmitter antenna*. The oscillating charges produce an electromagnetic field, a portion of which radiates into space. At the receiving end, the electric part of the electromagnetic field produces a current in another wire called the *receiver antenna*. This current is then detected by a suitable device.

If the charges that produce the electromagnetic wave are in a continuous back-and-forth oscillating motion, then the wave they produce will also oscillate con-

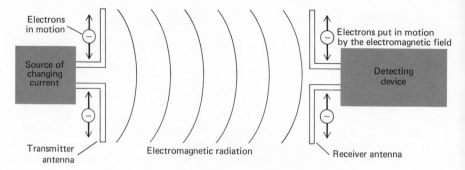

Figure 5-7 Communication with electromagnetic waves. Charges moving in the transmitter antenna emit electromagnetic radiation which produces a current in the receiver antenna.

tinuously. Such a wave carries very little information. Its detection does indicate that someone is transmitting, but since the signal does not change with time, no additional information is obtained. The electromagnetic wave can, however, be altered at the transmitter to carry information. Such an alteration is called *modulation*. The radiation can be simply turned on and off, producing pulses in the form of the Morse code. These pulses can be detected at the receiving end and decoded. There are much more sophisticated modulation techniques with which sound and visual information can be transmitted. We will discuss these later.

At the time of Hertz there were still many technical difficulties in the way of a practical wireless. The main problem was in the method of detection. Detecting the electromagnetic radiation with sparks as Hertz did is obviously very inconvenient. In addition, the production of a spark requires a strong field. An electromagnetic wave, just as a sound wave, attenuates with distance from the source. Although the dissipative losses are very small, the wave spreads and therefore the energy in a given region decreases as the distance from the source increases. At distances of more than a few yards from the source, the field is too weak to produce a spark.

In 1889 Oliver Lodge perfected a device for detecting radio-frequency electromagnetic signals which was much more sensitive than the earlier simple spark gap. The device is called a *coherer*. It consists of a spark gap inside a glass tube which is connected across a galvanometer and a battery. In the absence of radiation the gap is an insulator and current does not flow through it. Although the electromagnetic field may not be strong enough to produce a spark in the coherer, it does rip off some electrons from the atoms in the gap. Because of the free electrons, the gap becomes somewhat conductive and a small current from the battery flows through the circuit. This current is detected by the galvanometer. In 1894 with this device Lodge demonstrated signaling over a distance of 150 yards at the British Association meeting in Oxford. By this time there were a number of others who were actively experimenting with wireless communication. Most successful among these were Marconi in Italy and Popov in Russia.

Marchesi Guglielmo Marconi (1874–1937) was born in Bologna, Italy. His father was Italian and his mother Irish. He was educated at the University of Bologna and while there he became familiar with Hertz's experiments. He became very enthusiastic about the potentialities of this work and began to experiment with radio waves. By 1895 with an improved antenna and transmitter he was able to transmit radio signals through distances of over 2 kilometers. In 1896 he decided to go to England to obtain financial support for his experiments; and in 1897 he applied for and was granted the first patent for the practical application of Hertzian waves. Marconi's work in England was very successful. In 1897 he founded the Wireless Telegraph and Signal Company with a capital of £100,000. The company acquired patent rights all over the world and was very effectively organized for the future expansions in the field. He was able to hire a number of able scientists through whose work much of the later accomplishments became possible. The British Government was fully aware of the potential importance of wireless communications. They cooperated with Marconi and his company. In 1898 Marconi and Captain H.B. Jackson, who had also worked with wireless communication, performed an impressive demonstration. During naval maneuvers they communicated over a distance of 60 miles. They received wide acclaim and the future of wireless communication was assured.

During the next three years Marconi kept extending the range of communication and by 1900 he was ready to try the great test, wireless communication across the Atlantic. Success here would firmly establish wireless as a most important communication technique.

TRANSATLANTIC WIRELESS COMMUNICATION

When Marconi first proposed transatlantic wireless communication, there were many skeptics who believed that the project was impossible. They claimed that all electromagnetic radiation travels in a straight line and that therefore communication much beyond the horizon was not possible. But Marconi was confident in his project. He said, "These waves of mine will follow the earth." In fact both Marconi and his critics were right. Electromagnetic radiation does indeed travel in straight lines, but at high altitudes the earth is surrounded by a layer of ionized gas called the *ionosphere* which reflects the radio signals back to earth. The ionosphere is produced primarily by the action of the short wavelength components of the sun's radiation on the atoms in the upper atmosphere. This radiation produces ions by ripping electrons from the atoms. The ionosphere extends from 30 to 200 miles above the earth and for radiation at radio frequency it behaves as a mirror by reflecting the incident radiation. As a result radio communication over long distances is made possible (Figure 5-8). At the time of Marconi's experiments nothing was known of the ionosphere and therefore the skepticism of his critics was well founded.

For the transatlantic trial Marconi built the transmitter in Poldhu, Cornwall, and the receiving station in Cape Cod. His receiving antenna consisted of a huge, clumsy, inverted cone formed by 400 wires. Both the sending and the receiving aerials collapsed in a strong wind shortly after their construction. The sending station was rebuilt, but the receiving station at Cape Cod was abandoned and a new receiving antenna was built in Newfoundland. The electromagnetic signal was produced by sparks about 4 centimeters long and the receiver was an improved coherer with the current detected by telephone.

In describing this experiment Marconi wrote later:

We reached Newfoundland on December 6 [1901] and erected our signal station on Signal Hill. On December 12 and in spite of a raging gale we flew a kite carrying an aerial some 400 feet long. About 12:30 in the afternoon a succession of three faint clicks corresponding to the prearranged signal sounded

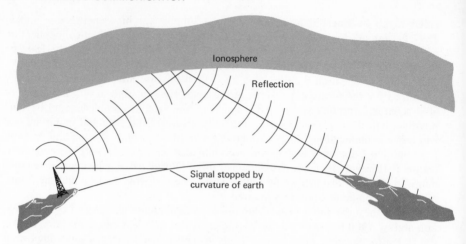

Figure 5-8 A radio signal propagated toward the ionosphere is reflected and can be detected beyond the line of sight of the antenna.

definitely and distinctly in the telephone held to my ear. This could only be that the electric waves sent out had traversed the 1700 miles of the Atlantic unimpeded.

The preparations for the experiments were done in secret and the announcement of its success was sensational. One of the first results of the success was that the AngloAmerican Telegraph Company threw Marconi out of Newfoundland claiming that he was infringing on their telegraph rights. The Canadian Government reinstated him in Glace Bay, Cape Breton Island. There were some people who claimed that the experiment was a fraud, but soon Marconi again communicated with England and this time his communication was confirmed by independent observers.

Marconi received many prizes and honors for his achievements. In 1909 at the age of 35 he was awarded the Nobel Prize in Physics. In retrospect it appears that Marconi's contributions to radio were overestimated. They were not as unique as the press releases of his time made them appear. At the same time and even before Marconi there were some people who were as advanced in radio communications as Marconi. For example, D.E. Hughes, who was a professor of music at a small college in Kentucky, came to England and in 1879 and in 1880 demonstrated wireless communications. Unfortunately, neither Hughes nor the people to whom he showed his experiments understood what was going on. On the other hand, A.S. Popov (1859-1906) in Russia did very competent work on wireless telegraphy for the Czarist Navy. By 1895 his technique was as advanced as that of Marconi. However, because of political unrest in Russia, his work did not continue as efficiently as that of Marconi. Nicholas Tesla, a Serbian scientist who immigrated to the United States, also worked on the wireless and made significant contributions to it.

Although Marconi's system was primitive and unreliable, the need for rapid communications was so great that it was very quickly put into use in shipping and military communications. Even before the transatlantic experiments some ships were being equipped for wireless. In 1899 the lightship *East Goodwin* was rammed and sunk. The crew was rescued because they were able to call for help with a wireless that had been installed the year before. The newspapers and the military were among the earliest users of the wireless. Already in 1904, during the Russo-Japanese War, news communications were transmitted to London by wireless telegraphy. The number of amateur operators also grew very rapidly. At first they only listened to Morse code conversations, but they were soon transmitting themselves.

Just as with the earlier telegraph, public interest in wireless was also excited in connection with a spectacular crime. In 1910 Dr. Hawley H. Crippen, an American physician who had been practicing in London, murdered his wife, buried her in a cellar, and escaped from England with his secretary on the liner *Montrose*. His secretary was disguised as a boy and they traveled as Mr. Robinson and son. The captain of the *Montrose*, George Kendall, had read about the crime and became suspicious of the Robinsons. The *Montrose* was one of the few ships which at that time was equipped with wireless and he telegraphed a message to Scotland Yard describing the couple. Scotland Yard dispatched Inspector Dews on a faster liner to intercept the *Montrose*. Before the *Montrose* reached port, Inspector Dews disguised as a pilot boarded the ship; he arrested Crippen and returned him to England.

The Marconi-type wireless communication system had many limitations. Most prominent among its drawbacks was poor signal reception. Although the coherer was an improvement on the spark gap, it was an unreliable and relatively insensitive device. The crystal detectors which soon replaced the coherer were somewhat better but still not entirely satisfactory. The signals received were very weak and techniques for amplifying them did not exist. Until about 1906 all communication was with Morse code and voice communication seemed remote.

A major breakthrough in radio technology came in 1904 with the development of thermionic tubes. That year the English physicist J.A. Fleming patented a two-element vacuum tube diode. The vacuum diode was an excellent detector of electromagnetic radiation and is still widely used. Two years after Fleming's invention Lee DeForest inserted another electrode into the diode configuration and produced the triode, a device of tremendous importance. With the triode it became possible to amplify and conveniently generate radio signals. Now developments in communication became so rapid that by 1907 Reginald A. Fesseden broadcasted speech over 200 miles of the eastern coast of the United States. Through the work of hundreds of engineers and inventors, but notably DeForest, Armstrong, Pupin, Meissner, Poulsen, and Fesseden, radio technology was developed to its full potential. In 1920 the first scheduled broadcasting was begun in Pittsburgh by station KBKA. By 1923 there were over 500 transmitters operating in the United States. They nearly all used the same wavelengths, which caused a great deal of confusion. In 1927 the Federal Radio Commission was formed and brought some order into broadcasting.

The development of broadcasting in Europe was nearly as rapid as in the United States. By 1931 there were 261 transmitting stations in Europe, again broadcasting over nearly identical wavelengths. This situation was brought somewhat under control by an international broadcasting conference in Prague.

Before we can discuss modern communication systems, we must explain the basic language of the field. This consists of a definition of units and basic concepts and a description of electric circuit components. These topics are discussed in Chapter 6.

6 Units, Definitions, and Circuit Components

Until the late 1500s Aristotelian thinking dominated western science. Within that framework it was sufficient to understand concepts in a plausible qualitative way. Experiments were very seldom performed to test theories or explanations and no attempt was made to establish ideas on a quantitative basis. Galileo was the first to challenge this approach. He realized the importance of being able to measure the relationship between cause and effect. This quantitative approach to science has had a most important effect on science and its applications. Through a quantitative approach it is possible to make exact predictions, which in turn test the understanding of the phenomenon. The ability to make quantitative predictions is essential for the building of devices that make up our technology. Beginning with Coulomb the studies of electricity and magnetism were conducted in the spirit of quantitative science.

Quantitative descriptions require units to measure the parameters and their interrelations. Any number of units may be used to measure a given entity; for example, it is just as valid to measure distance in yards as it is in meters. Over the years, however, some units have become more popular than others. It is not easy to obtain universal acceptance for the same set of units. Even in the simplest measurements, such as for weight or distance, dozens of units are in use. The metric system (abbreviated MKS) which was originated in 1901 is widely used especially in communication technology. In this system, distance is measured in *meters* (m), mass is measured in *kilograms* (k), and time is measured in *seconds* (s). All other physical quantities, such as force, energy, and power, can be expressed in terms of these three basic units. However, since to do this would be cumbersome, the most frequently used quantities have been assigned their own units. It is a rather interesting feature of the MKS system that with the exception of the three basic units, all units have been named after people who contributed to the concepts that the units describe. The relations between the MKS and the conventional British units are shown in Table 6-1.

Before we can discuss electrical units, we have to define three fundamental concepts: force, energy, and power. Our description will be brief but, hopefully, adequate. A more detailed treatment of these topics is found in the texts on elementary physics listed in the References.

ble 6-1   Relations between MKS and British Units

	MKS Units	British Units	Conversion
Time	Second (s)	Second	
Distance	Meters (m)	Feet	1 meter = 3.28 feet
Mass	Kilograms (k)	Slugs	1 kilogram = 0.0685 slug
Force	Newtons	Pounds (lb)	1 newton = 0.225 pound
Energy	Joules	Foot-pounds	1 joule = 0.738 foot-pound
Power	Watts	Foot-pounds/min	1 watt = 44.25 foot-pounds/min
		Horsepower	1 watt = 0.00134 horsepower

NITS AND DEFINITIONS

orce

In the MKS system of units force is measured in *newtons*. In terms of the basic units the newton is in units of (kilogram × meter)/second². This can be obtained from Newton's law which relates force to mass and acceleration:

$$F = ma$$

Here m is the mass of the object measured in kilograms and a is the acceleration of the object measured in meters per second per second (meter/second²). Thus 1 newton is the force required to accelerate a 1-kilogram mass at a rate of 1 meter per second per second. By substituting these units into Newton's equation, we obtain the unit for the force in the MKS system.

Of course, we can express force just as well in the British system of units in which it is measured in pounds. The relationship between the pound and the newton shown in Table 6-1 is derived as follows. A 1-pound force will accelerate a 1-slug mass at a rate of 1 foot per second², that is,

$$1 \text{ pound} = 1 \text{ slug} \times 1 \text{ foot per second}^2$$

but

$$1 \text{ slug} = 14.59 \text{ kilograms} \quad \text{and} \quad 1 \text{ foot} = 0.3048 \text{ meter}$$

therefore

$$1 \text{ pound} = 14.6 \text{ kilograms} \times 0.305 \text{ meter per second}^2$$
$$= 4.45 \text{ kilogram-meters per second}^2$$
$$= 4.45 \text{ newtons}$$
$$1 \text{ newton} = 0.225 \text{ pound}$$

Energy and Work

Energy is a very important concept in science. We find reference to energy in connection with widely different phenomena. We speak of atomic energy, heat energy, solar energy, chemical energy, kinetic energy; we even speak of people as being "full of energy." The common factor that ties all these manifestations together is the possibility of obtaining *work* from all these sources. In the words of Maxwell, "Energy is the capacity of doing work." But what is "work"? We all have a qualitative idea of what the word means, but in science we need a quanti tative definition. In its simplest form work is defined in terms of a force actir on a body through a distance. Algebraically this is expressed as

$$W = F \times D$$

where W is the work produced by a force F which acts through a distance D (Figur 6-1). We can see that in the MKS system work is measured in units of newton meters. The unit representing the newton-meter is called the *joule*, after the Englis physician James Prescott Joule (1 joule = 1 newton-meter). To give a numerica example, when a force of 2 newtons acts on a body through a distance of 1 mete the work done is equal to 2 joules. The connection between energy and work simple: energy is required to do work and energy is measured in the same unit as work. In fact there is a one-to-one correspondence between them; it take 2 joules of energy to do 2 joules of work.

To illustrate the conversion of energy to work let us consider a container c burning gasoline. During the process of burning, chemical energy is released a the gasoline molecules rapidly combine with oxygen. This energy generates force that ruptures the container walls and sets the fragments into rapid motion Because the force acts through a distance, work is done on the fragments. Bu because of their motion, these moving fragments can in turn do work, for ex ample, by hitting other objects and putting them into motion. Therefore we con clude that a moving object possesses energy, which is called *kinetic energy*. Th process does not end here: through friction the fragments are slowed down an the kinetic energy is converted into heat energy, which in turn can be converte into other forms of energy. Here lies the importance of the energy concept: i all physical processes energy is conserved; that is, through work one form o energy can be converted into another, but the total amount of energy remains unchanged.

Power

The amount of work done, or energy expended, per unit time is called *power*. The algebraic expression for power is

$$P = \frac{E}{\Delta t}$$

where E is the energy expended in a time interval Δt.[1] The unit of power is the *watt*, named after the Scottish engineer James Watt. One watt is equal to 1 joule per second. For example, in a 100-watt lightbulb, 100 joules of electric energy are converted into heat and light every second.

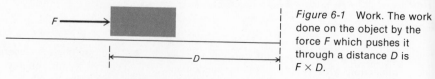

Figure 6-1 Work. The work done on the object by the force F which pushes it through a distance D is $F \times D$.

[1]The Greek letter Δ (delta) is used to symbolize an interval or a change in an entity.

Electrical Units

The units describing the various electrical quantities can all be defined in terms of the fundamental MKS units. However, this again would be too cumbersome. Therefore the frequently used electrical quantities are given separate designations. We shall start our description of electrical units by defining the electric charge.

Electric Charge

We recall that Coulomb found the force between two charged bodies to be proportional to the product of their charges and inversely proportional to the square of the distance between them. Mathematically this is expressed by Coulomb's law as

$$F = \frac{KQ_AQ_B}{R^2}$$

where R is the distance between the two bodies with charges Q_A and Q_B. As we mentioned in the discussion of Coulomb's law, K is the constant of proportionality and depends on the set of units used. In the MKS system electric charge is measured in units called *coulombs* and the constant or proportionality is 9×10^9. In the MKS units Coulomb's law is

$$F = \frac{9 \times 10^9 Q_AQ_B}{R^2}$$

From this equation the force between two charges of 1 coulomb each separated by 1 meter is 9×10^9 newton, or about 1 million tons. A coulomb represents a very large amount of charge. For example, it takes the charge of 6×10^{18} electrons (6 billion billion electrons) to make 1 coulomb of charge.

Electric Current

An electric current is produced by the motion or flow of charges. The strength of the current depends on the amount of charge flowing past a given point in a given period of time. In the MKS system current is measured in *amperes*. One ampere is 1 coulomb of charge flowing past a point in 1 second. Although 1 coulomb is a very large charge, it is not difficult to obtain a 1-ampere current. The current flowing in a light bulb is on that order.

Potential Difference or Voltage

The electrons, which are the charge carriers in the conducting material, are not entirely free moving. Depending on the nature of the conductor, the electrons are more or less bound to the atoms and their motion is further impeded by collisions with the atoms within the conductor. This opposition to the motion of charges has to be overcome if the current is to flow. In other words, a force must be applied to the electrons in order to maintain their directed motion through conductor. Force is exerted on charges by an electric field. Therefore in orde for current to flow between two points there must be an electric field betwee them. The current flows as a result of the force exerted on the charges by th electric field. The electric field can be the result of a difference in charge con centrations between two points, as in a voltaic cell, or it can be produced by changing magnetic field, as in a generator. In the MKS system the electric fiel is measured in units of volt per meter. The force on a charge Q is then given b the product of the charge and the electric field E. Algebraically this is represente by $F = EQ$. Thus a force of 1 newton is exerted on a 1-coulomb charge by a 1-vol per-meter field. Because there is a force applied to the charge and the charge i moving, there is work done on it. This work is the product of the force F and th distance D through which the electric field acts on the charge (Figure 6-2):

$$\text{work done on the charge} = EQD = EDQ \qquad (6\text{-}1$$

Since D is measured in meters, the unit of the product ED is volts. This produc of the electric field and the distance over which it acts is an important paramete in practical application. It is called the *voltage* or *potential difference* and is usuall symbolized by the letter V. Using this symbol Equation (6-1) is written as

$$\text{work done on the charge} = QV$$

The voltage between two points is a measure of the work done on or the energ given to the charge as it moves between the points. Some of this energy is alway dissipated through collisions of electrons with atoms. Here electric energy is con verted into heat in a process that is analogous to frictional dissipation. With suitabl devices electric energy can also be converted into mechanical and chemical energy

We will now illustrate these concepts by an example. Let us consider the 6-vol car battery shown in Figure 6-3. It has a postive (+) terminal and a negative (− terminal. Through chemical reaction an excess of electrons has built up at th negative terminal which produces an electric field between the two terminals The label on the battery says 6 volts, so we know that this electric field time the distance between the terminals is 6 volts. The excess electrons can move from the negative to the positive terminal only if the terminals are connected througt a path that can conduct electricity. The repulsion of the electrons from the nega tive terminal causes energy to be stored in the battery. This energy is releasec as the electrons move from the negative terminal to the positive terminal. Because the voltage between the terminals is 6 volts, for every coulomb of charge trans ferred, 6 joules of energy are released from the battery. The amount of energy released by the battery is independent of the path of the charge between the ter minals. The electric field will follow the conducting path. In this sense the situation is closely analogous to gravitational potential energy. An object on top of the

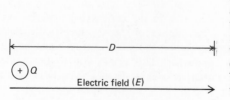

Figure 6-2   Work done by an electric field. The electric field E exerts a force EQ on the charge Q. Under the action of this force the charge moves a distance D. Therefore the work done on the charge is EQD.

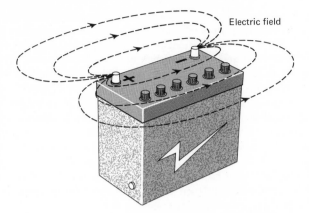

Figure 6-3 The electric field lines in air surrounding the terminals of a battery.

hill has energy by virtue of its position. As the object rolls down the hill, the potential energy is converted into kinetic energy. The amount of kinetic energy gained by the object does not depend on the detailed path, it depends only on the net height from which the object falls (Figure 6-4).

The disposition of the energy released by the battery depends on the nature of the conducting path. If the conducting path is through a light bulb, all the energy is used to heat the filament of the bulb. On the other hand, if the current flows through the windings of an electric motor, the electric energy is converted into mechanical energy.

Finally, we must remember that if there is a voltage between two points, then there will be a force exerted on a charge placed in the region between these points. If the charge is negative (an electron), then the force will move the charge away from the negative and toward the positive point.

Electrical Resistance

As we have already mentioned there is in all materials an opposition to the flow of current. This property is called the *electrical resistance* of the material or simply

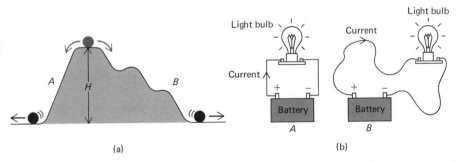

Figure 6-4 (a) The amount of energy released as the object rolls down the hill depends only on the net change in the altitude *H* of the object. Thus the amount of released energy through paths *A* and *B* is the same. (b) Similarly for a given current the amount of energy obtained from a battery depends only on its voltage and is independent of the current path. Thus the same amount of energy is released through paths *A* and *B*.

resistance. Resistance is measured in *ohms* after the German scientist Georg S.
Ohm (1787–1854) who studied the phenomenon.

There are great differences in the resistances of various materials. In com
paring materials of the same size and shape, carbon has a resistance about 2000
times greater than that of copper, and glass about 100 billion times higher than
that of carbon. The current that can actually pass through materials such as glass
is so small that these materials are considered insulators of electricity. The re
sistance of the material depends not only on its composition but also on its shape
The smaller the area through which the current has to pass, the larger the resistance
to the current flow. And as we would expect the resistance increases with the
length of the current path.

Given a voltage source such as a battery or generator, the amount of current
that flows between its terminals depends on the resistance of the material that
connects them. The higher the resistance, the smaller the current flow. If the
voltage of the source is increased, the current also increases. In 1826 Georg Ohm
found a simple relationship that connects voltage current and resistance. The
connecting equation (which is now called *Ohm's law*) is

$$V = IR \quad \text{or} \quad I = \frac{V}{R}$$

Here R is the resistance of the material connected across the terminals of the
voltage source. Knowing the value of any two of the quantities, the third can be
computed from this law. For example, if a 10-ohm resistance is connected across
a 100-volt source, a 10-ampere current will flow from the source.

$$I = \frac{100 \text{ volts}}{10 \text{ ohms}} = 10 \text{ amperes}$$

Conversely, if we know that the current through a 10-ohm resistor is 10 amperes,
then we can compute that the voltage across the resistor is 100 volts.

$$V = 10 \text{ amperes} \times 10 \text{ ohms} = 100 \text{ volts}$$

Electric Power

We have already shown that the work done on the charge Q transported from
one terminal to the other by a voltage V is QV. This is also the energy supplied
by the voltage source. In order to obtain an expression for the power P delivered
by the source, we divide the energy by the time interval Δt over which this energy
is being supplied. Therefore

$$P = \frac{QV}{\Delta t}$$

But Δt is also the time interval during which the quantity of charge Q is moved
across the terminals of the voltage source. Therefore $Q/\Delta t$ is simply the current
through the connection. Substituting I for $Q/\Delta t$, we obtain

$$P = VI$$

Thus the power delivered by a voltage source is the product of the voltage and
current. If the current from a 10-volt source is 5 amperes, the power delivered is
50 watts:

$$P = 10 \text{ volts} \times 5 \text{ amperes} = 50 \text{ watts}$$

Using the relationships given by Ohm's equation we can write two alternate and, of course, equivalent expressions for power:

$$P = I^2R \qquad \text{and} \qquad P = \frac{V^2}{R}$$

We shall illustrate the use of Ohm's law and the power equations by examining the operating condition for a 100-watt light bulb. The wattage of the bulb is specified; therefore we know that in the bulb 100 watts of electric power are converted into heat and light. This power is supplied by the 110-volt supply provided by the electric company. The resistance of the light bulb can be obtained from the equation for power:

$$P = \frac{V^2}{R}, \qquad R = \frac{V^2}{P} \qquad \text{or} \qquad R = \frac{(110)^2}{100} = 121 \text{ ohms}$$

For the bulb to operate efficiently the resistance of the wires that connect the bulb to the socket must be much lower than the bulb resistance. That is, of course, the case. The resistance of 1000 feet of typical household electric wire is about 2 ohms. The current through the light bulb is calculated from Ohm's law

$$I = \frac{V}{R} = \frac{110}{121} = 0.91 \text{ ampere}$$

The resistance and current of the light bulb that we calculated are the values during steady operating conditions. The situation is different when the bulb is first turned on. The resistance of most materials increases as their temperature is raised. This is due to the fact that random motion of both the electrons and atoms in the material increases with temperature. This random motion impedes the directed flow of electrons that constitute the current. The resistance of a light bulb increases by about a factor of 10 when the bulb is heated to its operating temperature. Thus the resistance of the cold bulb is only about 10 ohms and therefore the initial current is 10 times higher than the operating current. Light bulbs burn out most frequently when they are first turned on. The reason for this is the large initial current. Once the filament is weakened, it tends to break under the stress of this current surge.

ELECTRIC CIRCUITS

An electric circuit consists of electric components connected across a voltage source. In this section we shall describe three basic circuit components: resistor, capacitor, and inductor. They are used in various combinations to control the currents and voltages in a circuit. Photographs of some commercial circuit components are shown in Figure 6-5.

Resistor

The resistor is the simplest electric component. Its electrical property is to oppose the flow of current. A simple electric circuit consisting of a battery and a resistor is shown symbolically in Figure 6-6. The current through the circuit is

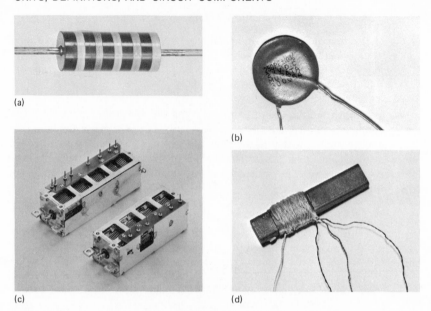

Figure 6-5 Photographs of (a) commercial resistor, (b) fixed capacitor, (c) variable capacitor, and (d) inductor. [(a) Courtesy of Allen-Bradley Company; (b) and (d) courtesy of Toshiba Electronic Corporation; (c) courtesy of Hammarlund.]

determined by Ohm's law. In this case the voltage of the battery is 6 volts and the resistance is 12 ohms; therefore the current through the circuit is ½ ampere. A more complicated resistive circuit is shown in Figure 6-7a. Here three resistors are connected across a battery. The total resistance in the circuit is the sum of the three resistances, which in this case is 6 ohms. Because the source is a 12-volt battery, the current flowing through the circuit is 2 amperes. The same current flows through each resistor, producing a voltage that is determined by Ohm's law. The voltage across resistors R_1, R_2, and R_3 is 2 volts, 4 volts, and 6 volts, respectively. The sum of the voltages across the resistors is equal to the source voltage. This is always the case in a series resistive circuit. Electric circuits can be highly complex as is shown in Figure 6-7b. Here again the currents and voltages of the individual components are obtained basically from Ohm's law, but the detailed treatment is outside the scope of our discussion here.

The light bulb is one example of a primarily resistive electric component. Here the resistance is used to convert electric energy into heat and light. In most circuit applications resistors are used to control the current through the circuit. For this purpose resistors are manufactured with a wide variety of resistance values. By

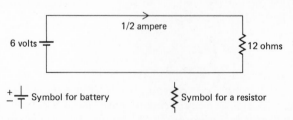

Figure 6-6 A simple resistive electric circuit. The arrow indicates the direction of conventional current. (The direction of electron flow is opposite to this.)

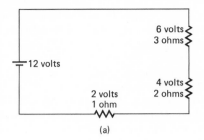

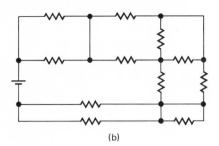

(a)   (b)

Figure 6-7 Resistive circuits.

controlling the composition of the resistive material, resistors can be made with values anywhere from a fraction of one ohm to millions of ohms.

Capacitor

The capacitor is a circuit element that stores electric charges. In its most elementary form a capacitor consists of two parallel conducting plates separated by an insulator, such as air, glass, or mica (Figure 6-8). When the plates are connected across the terminals of a battery, electrons start to flow away from the negative terminal, but the charges cannot cross the insulator between the plates. Therefore the electrons "pile up" at the plate connected to the negative terminal. The excess negative charges produce an electric field which causes a repulsion of electrons from the positive plate. If the voltage across the capacitor is increased, the stored charge in the capacitor also increases. The relation between the charge and the voltage across the capacitor is given by $Q = CV$. Here C, the *capacitance,* is the measure of the charge capacity of the capacitor. It is measured in *farads*. As is evident from this equation the larger the capacitance, the more charge is stored for a given voltage. For example, with a 1-volt source a 1-farad capacitor stores 1 coulomb of charge, whereas a $\frac{1}{2}$-farad capacitor stores $\frac{1}{2}$-coulomb. In fact a 1-farad capacitance is too large for practical applications. Most capacitors are on the order of 1-millionth farad or less.

If the plates of a charged capacitor are connected through a conductor, electrons flow from the negative to the positive plate. The current, of course, flows only as long as there is a charge difference between the plates. During this time the capacitor behaves as a battery. We conclude that a charged capacitor stores electric energy.

Capacitors are manufactured in various sizes (Figure 6-5b). They can also be constructed so that the capacitance is variable (Figure 6-5c). In the variable capacitor shown in the figure the capacitance is varied by changing the overlap of the surface

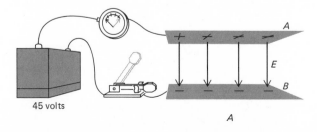

45 volts

A

Figure 6-8 A simple capacitor.

plates. Because the amount of stored charge is proportional to the area of the opposing plates, changing the effective capacitor area changes the capacitance.

Inductor

Perhaps the easiest way to understand the operation of an inductor is to examine the behavior of a simple circuit such as shown in Figure 6-9. With the circuit closed by the switch, the current flowing through the conductor is given by Ohm's law, $I = V/R$, where R is the resistance of the conductor. This, however, is not the current that flows immediately after the closing of the switch. In fact, as shown in Figure 6-9, the current starts from zero and builds up to its final value in a finite length of time. To see how this comes about, let us examine what happens immediately after the switch is closed. As the switch is closed a current begins to flow and with it a magnetic field starts growing around the conductor (Figure 6-10). This growing magnetic field produces an electric field which, in accord with the Lenz law, opposes the flow of charges in the conductor. The build up of current is therefore slowed down. When the current reaches the final value, the magnetic field is no longer changing and the counterelectric field disappears (Figure 6-10b). Similarly, if the switch is opened while a current is flowing through the circuit, the diminishing magnetic field produces an electric field that aids the current flow and causes it to decrease gradually. This property of the conductor to produce a countervoltage due to a changing current is called *inductance*. Inductance is symbolized by the letter L and it is measured using a unit called the *henry*. Quantitatively inductance is defined by the equation

$$L = \frac{V}{\Delta I/\Delta t}$$

where $\Delta I/\Delta t$ is the change in the current ΔI during the time interval Δt and V is the countervoltage produced by the changing current. The countervoltage is the product of the electric field and the distance over which the field is produced. Thus a 1-volt countervoltage is produced by a 1-henry inductor with a current that changes at a rate of 1 ampere per second.

The inductance of an ordinary conductor is very small, so that the time required for the current to build up or decay is less than 1-millionth second. However, the inductance can be increased by winding the conductor into a coil. Since the magnetic fields of the individual turns are cumulative, the countervoltage produced by a given current change is proportionally increased, resulting in a higher inductance. A still larger inductance can be obtained by putting a soft iron core inside the coil.

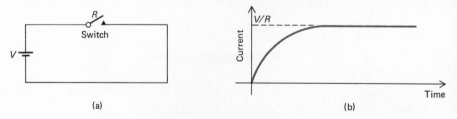

Figure 6-9 Inductance. The final current in the circuit (a) is V/R but because of the inductance, this value is established only after a finite length of time (b).

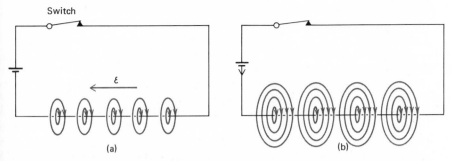

Figure 6-10   Induction. When the current first starts flowing, an electric field is generated by the growing magnetic field. (a) This electric field opposes the current flow. (b) When the current reaches its final value, the associated magnetic field stops growing and the electric field vanishes.

Circuit elements designed specifically to have a given inductance are called *inductors* (Figure 6-5c). They are used primarily to reduce rapid current fluctuations and in conjunction with capacitors to construct resonant circuits.

Resonant Circuit

One of the most important circuits in communication applications is the *resonant circuit* (sometimes called a *tuned circuit*). In its simplest form it consists of the capacitor-inductor combination shown in Figure 6-11a. We will explain the operation of the resonant circuit with the aid of Figure 6-11. Let us assume that initially the capacitor is charged and the inductor is connected across it (Figure 6-11a). The capacitor begins to discharge, but the discharge is not instantaneous because the

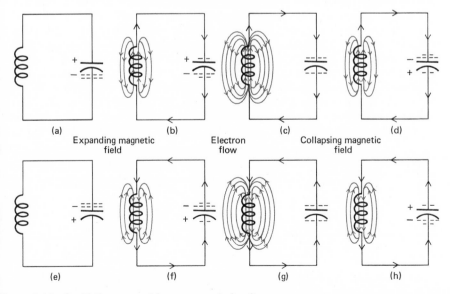

Figure 6-11   Oscillating current in a resonant circuit.

countervoltage produced by the growing magnetic field around the inductor retard
the current flow. The capacitor discharges at a rate determined by the capacitance
and the inductance of the circuit. At the instant when the capacitor is discharged, the
current stops and the magnetic field around the inductor starts collapsing (Figure
6-11c). This produces a voltage across the inductor which causes the current to
continue to flow in the same direction. Because of the continuing current, the ca
pacitor again becomes charged but this time in the opposite direction (Figure 6-11e)
Electrons now start flowing in the opposite direction from the upper to the lowe
plate and the whole procedure is reversed. The current then continues to oscillate
back and forth as shown in Figure 6-12a. In an actual circuit, there is always some
resistance which causes the oscillating current to dissipate as shown in Figure 6-12b
The frequency of the oscillating current, that is, the number of times the current
reverses direction per second, is uniquely determined by the circuit capacitance
and inductance.

It has been shown mathematically and confirmed experimentally that the fre
quency of a resonant circuit is given by

$$f = \frac{1}{2\pi\sqrt{LC}}$$

Thus, for example, if the capacitor is 1-millionth farad (10^{-6}) and the inductor i
1-thousandth henry, the frequency is

$$f = \frac{1}{2\pi\sqrt{10^{-6} \times 10^{-3}}} = 5000 \text{ cycles per second}$$

The electric resonant circuit is very closely analogous to a mechanically resonant
system such as a swing, a pendulum, or a tuning fork. When a swing is displaced from
its vertical position, it will oscillate back and forth just as the charges on the plates
of the capacitor. The resonant frequency of oscillation is uniquely determined by
the parameters of the mechanical system. In the case of a swing the length of the
rope and the gravitational force determine the resonant frequency. A swing can be

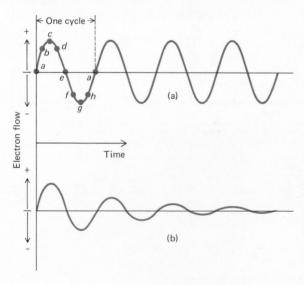

Figure 6-12   Flow of electrons in a resonant circuit.
(a) Theoretical flow, (b)
actual flow.

put into motion by any force, but by far the largest swing amplitude is obtained when the force on the swing is periodic at the resonant frequency of the swing. Similarly, any voltage applied across the resonant circuit will cause a current to flow through it, but the current will be largest when the frequency of the voltage is at the circuit resonant frequency. In communication technology, resonant circuits are most commonly used to generate electric signals at a specific frequency and to detect signals at a selected frequency. These applications will be discussed in greater detail.

Having concluded the explanation of fundamentals, we are now ready to continue our discussion of radio.

SUMMARY

FORCE (F)   When a force is applied to a body of mass m, the body will accelerate. The motion of the body is governed by Newton's equation $F = ma$, where a is the acceleration.

ENERGY AND WORK (W)   Energy is the capacity of doing work. When a force F acts on a body through a distance D, the work done on the body is $W = F \times D$.

POWER (P)   Power is the amount of work performed per unit time.

ELECTRIC CHARGE (Q)   Electric charge is measured in units of coulomb. The force between two bodies with charges Q_A and Q_B separated by a distance R is given by Coulomb's law, which in MKS units is

$$F = \frac{9 \times 10^9 \, Q_A Q_B}{R^2}$$

ELECTRIC CURRENT (I)   An electric current is a flow of electric charges. Current is measured in amperes. One ampere is 1 coulomb of charge flowing past a point in 1 second.

POTENTIAL DIFFERENCE OR VOLTAGE (V)   The voltage between two points is the measure of work W done on a charge Q when it is moved between the two points. The work is $W = QV$. If there is a voltage between two points, then a force is exerted on a charge located in the region between the two points.

RESISTANCE (R)   Resistance is the measure of the opposition to current flow. Resistance is measured in units of ohm. The relationship between current voltage and resistance is given in Ohm's law, which is $V = IR$.

ELECTRIC POWER   The electric power in units of watt is given by the following relationships: $P = I^2 R = V^2/R = VI$.

ELECTRIC CIRCUIT   An electric circuit consists of electric components connected to a voltage source.

RESISTOR   The resistor is a circuit component that resists current flow.

CAPACITOR   The capacitor is a circuit component that stores electric charge.

INDUCTOR   The inductor is a circuit component that opposes changes in current flow.

RESONANT CIRCUIT   The resonant circuit is a combination of a capacitor and inductor. The current flow in the circuit is largest when the frequency of the current is at the resonant frequency of the circuit.

7 Modern Radio

The two main problems that had to be solved before practical radio voice communication could be established were the detection and the amplification of electromagnetic signals. Both of these problems were solved by the development of the
vacuum tube diode and *triode*. In the following sections we shall discuss the operation and application of these devices.

DETECTION OF RADIO WAVES

Let us first consider the problem of detecting a radio signal. The radio waves im-
pinge on the receiving antenna and produce in it a current that is proportional to
the electromagnetic waves. The frequency of the antenna signal current, that is, the
number of times the electrons change their direction, is the same as the frequency of
the radio waves. However, this frequency is so high (thousands of cycles per second)
that mechanical devices such as galvanometers or speakers cannot respond to it.
The basic problem of radio signal detection is the conversion of the high-frequency
signal into one that can drive devices whose output is detectable by our senses (these
devices are called *transducers*). As a specific illustration of the detection process,
let us consider again the wireless transmission of the Morse code (Figure 7-1a).
In the figure we show schematically the wave train generated in the transmission
of the letter "n" (-·). During each Morse code pulse, the electric field of the radio
signal changes direction many times. For example, if the pulse lasts 0.1 second and
the radio frequency is 100,000 cycles per second, the electric field reverses direction
10,000 times during one pulse period. The current in the receiving antenna has the
same shape as the radio signal that produced it (Figure 7-1b). However, our transducers (speakers, galvanometers) cannot respond to radio frequencies and, in fact,
we do not want them to respond to these frequencies because there is no information
in the frequency of the radio wave. The message is in the duration of the code pulses.
It is possible to detect the pulses with a coherer because the coherer conducts current only during the interval when the radio frequency pulse is passing through it.
The current through the circuit containing the coherer flows in pulses that have the

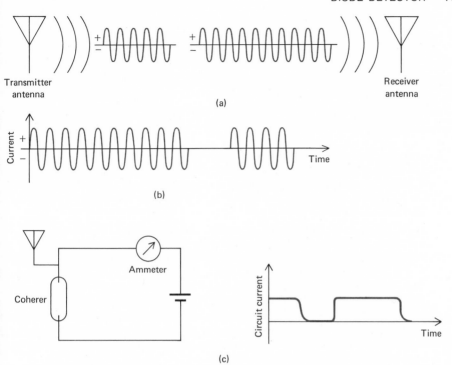

Transmitter
antenna

Receiver
antenna

(a)

(b)

Ammeter

Coherer

(c)

igure 7-1   Wireless transmission of Morse code. (a) Schematic representation of the radio
waves transmitting the Morse code letter "n." (b) The current in the receiver antenna
produced by the radio waves. (c) A coherer circuit and the current flowing through it.
This current is detected by the ammeter.

same duration as the code pulses (Figure 7-1c). These current pulses can then drive
an ammeter or some other device. Detection with a diode is similar, but much more
versatile and reliable. Before we explain the physics of the diode, we will describe
its functional operation.

DIODE DETECTOR

The symbolic representation of the vacuum tube diode is shown in Figure 7-2a. As
shown in the diagram, the vacuum tube diode consists of two elements, the *anode*
(or plate) and the *cathode,* sealed into an evacuated envelope. The operation of the
diode is based on one simple property: Current can flow through the diode only when
the voltage on the anode is positive with respect to the cathode (Figure 7-2b). When
the voltage on the anode is negative, the diode behaves as an insulator and no current
flows through it (Figure 7-2c). The use of the diode in the detection of radio signals
is shown in Figure 7-3a. The signal is again assumed to be the Morse code letter
"n." The antenna is connected through the primary winding of a transformer to
ground. As before, the current flowing in the antenna (and therefore also in the primary coil of the transformer) is identical in form to the radio signal (Figure 7-3b).
This primary current induces a corresponding voltage in the secondary transformer
coil; and here begins the operation of the diode. Although the secondary voltage has

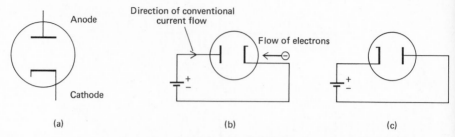

Figure 7-2 Vacuum diode. (a) Schematic representation of a diode. (b) Current flows through the diode when the anode is positive with respect to the cathode. (c) Current does not flow through the diode when the anode is negative.

the same shape as the radio signal, the current through the secondary does not. Because of the diode, the secondary current flows only during the intervals in which the anode is positive. This current therefore always flows in the same direction and consists of pulses as shown in Figure 7-3c. The voltage across resistor R (which can be obtained from Ohm's law, $V_R = IR$) has the same shape as the current through it. This process, in which an alternating signal (one that changes direction) is converted into a direct signal (one that does not change direction), is called *rectification*. Before the detection circuit is complete, a capacitor has to be connected across the resistor R as shown by the broken lines in Figure 7-3a. The capacitor smooths out the rectified voltage pulses into a signal that is a replica of the transmitted Morse code (Figure 7-3d).

The smoothing action of the capacitor can be explained with the aid of Figure 7-4. The first pulse of current in the pulse train produces a voltage across the resistor R and charges the capacitor to the same voltage. Without a capacitor, the voltage across the resistor follows the current shape exactly, and so, when the current

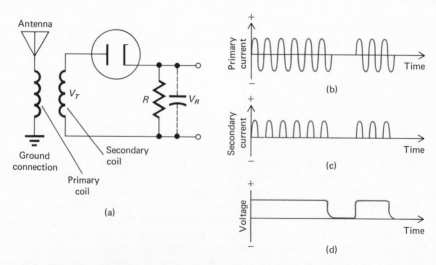

Figure 7-3 Detection of radio signals. (a) Detection circuit using a diode. (b) The current in the antenna and the primary winding. The shape of this current is identical to the radio signal. (c) The shape of the current through the diode and the voltage across the resistor. (d) Voltage across the capacitor resistor combination. This is the desired detected signal.

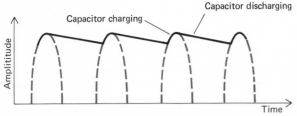

Capacitor discharging

Capacitor charging

Figure 7-4 The smoothing action of a capacitor. The solid line shows the voltage across the capacitor and the broken line shows the current through the diode.

decreases, the voltage across the resistor also drops. But with the charged capacitor connected across the resistor, the voltage is maintained at nearly peak value during the off-current periods. In fact, some discharging of the capacitor does occur through the resistor and this does cause a small decrease in the voltage. The capacitor, however, is recharged again during each current pulse and thus the voltage variations are small.

VACUUM TUBE DIODE

In 1889 while experimenting with light bulbs, Thomas Edison sealed a metal plate inside an evacuated bulb containing the filament. He noticed that when he connected the plate to the positive terminal and the filament to the negative terminal of a battery, a current flowed in the circuit (Figure 7-5). Edison did not consider this effect significant and did not investigate it any further. The Edison effect was correctly explained in 1904 by the English physicist Sir John Fleming who built and patented the first diode.

The operation of the vacuum tube diode is simple. In addition to the anode and the cathode, the tube also contains a filament which heats the cathode. When the cathode is heated, some of the electrons in the cathode material gain sufficient energy to overcome the internal atomic binding forces and escape from the cathode surface (Figure 7-6a). If the anode is made positive with respect to the cathode, the emitted electrons are attracted to the anode and therefore a current flows from the cathode to the anode. Since electrons are continuously supplied by the heated cathode, the diode serves as an effective conductor of electrons from the cathode to the anode (Figure 7-6b). If, on the other hand, the cathode is made positive, the emitted electrons are pulled back to it and the space between the cathode and the anode is left empty (Figure 7-6c). Since the anode cannot emit electrons, a current does not flow. Thus the diode behaves as a very good conductor when the anode is positive with respect to the cathode and as an insulator when the anode is negative with respect to the cathode.

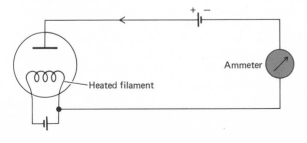

Ammeter

Heated filament

Figure 7-5 The Edison effect. When the plate is connected to the positive terminal of a battery, a current flows in the circuit (arrow shows the direction of conventional current flow.

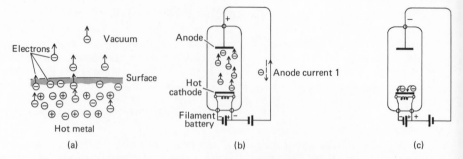

Figure 7-6 The operation of a diode. (a) Emission of electrons from a hot-cathode surface (b) When the anode of the diode is positive the emitted electrons flow from the cathode to the anode. (c) When the anode is negative, the emitted electrons are forced back toward the cathode and current cannot flow through the tube.

Commercial diodes are cylindrically symmetric. The anode, in the form of a sleeve, surrounds the heater and the cathode, which is usually coated with an oxide of strontium or barium to increase its electron emission. Electrical connection to the elements inside the vacuum envelope is made with leads that are sealed into the envelope. These leads are plugged into a socket that is wired into the appropriate circuit. A vacuum tube diode is about the size of a light bulb. Vacuum tube diodes were so superior to other detectors that by 1905 they were widely used in radio communication.

TRIODE AMPLIFIER

Although the diode solved the problem of detection, there was still no way of amplifying the detected signal. This was the great obstacle in long-distance communication since in many situations the detected signals were not strong enough to drive the transducers.

Electronic amplification was made possible by the invention of the triode in 1906 by Lee DeForest. DeForest reasoned that if the diode made such a wonderful device, surely some improvement could be obtained by adding a third electrode to this configuration. The resulting device, which is now called the *triode*, is one of the most important inventions of modern technology.

A schematic diagram of the triode is shown in Figure 7-7a. In its basic construction the triode is similar to the diode except that a third electrode, called a *grid*, is inserted between the plate and the cathode. The grid is placed close to the cathode. It is made of a metal mesh so that electrons can flow through it from the cathode to the anode. In the normal triode operation the anode is made positive and the grid is made negative with respect to the cathode. Since the grid is negative it repels the electrons; some of them, however, do pass through the grid and reach the anode. Because the grid is close to the cathode, it has a great influence on the amount of current that flows from the cathode to the anode. Small voltage changes on the grid produce large changes in the current flowing through the tube (Figure 7-7b). If the voltage on the grid is made more positive, the tube current increases and conversely, if the voltage on the grid is made more negative, the current decreases. The control of the current flowing through the tube by the grid results in signal amplification.

The circuit diagram for a simple triode amplifier is shown in Figure 7-8a. The anode is made positive and the grid negative with respect to the cathode by appropriate connections to batteries or power supplies. These voltages are called the *bias* for the tube. In the circuit shown in Figure 7-8a, a bias voltage between the anode and the plate is 200 volts and between the grid and the cathode it is −3 volts. With these voltages the current flowing through the triode is typically about 5 milliamperes (1 milliampere is 1/1000 ampere). This current also flows through the 20,000-ohm resistor that is connected in the anode circuit. The voltage across the resistor is calculated from Ohm's law and is

$$V = IR = \frac{5}{1000} \text{ ampere} \times 20,000 \text{ ohms} = 100 \text{ volts}$$

The construction of the triode is such that a change in voltage of 1 volt on the grid typically results in a current change of 1 milliampere. For example, when the voltage between the cathode and grid changes from −3 to −2 volts, the current through the triode increases from 5 to 6 milliamperes. This in turn changes the voltage across the resistor from 100 volts to 120 volts, that is, a 20-volt change. Thus a 1-volt change in the grid voltage produces a 20-volt change across the resistor. In other words, through the triode the change in the voltage has been amplified by a factor of 20.

In an amplifier the signal source (which may be the output of an antenna) is connected in series with the bias supply into the grid circuit. As the signal voltage varies (Figure 7-8b) it adds or subtracts from the bias voltage so that the net voltage on the grid is the algebraic sum of the bias and the signal voltages (Figure 7-8c). The changing grid voltage produces a changing anode current (Figure 7-8d), which in turn results in a changing voltage across the resistor. The voltage variations across the resistor are an amplified replica of the signal. This changing voltage across the resistor is called the *output* of the amplifier.

So far we have discussed only the amplification of voltage by the triode circuit. The circuit, however, also amplifies the input power; this is an important point. The signal voltage alone could be amplified by a transformer, but then the current would decrease in proportion to the increase in the voltage and the power of the signal would remain unchanged. Since the ability of a signal to drive a speaker or other transducers depends on its power, the amplification of signal power is impor-

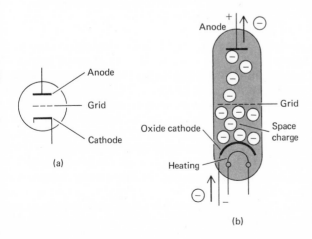

(a)

Anode

Grid

Cathode

Anode

Oxide cathode

Heating

Grid

Space charge

(b)

Figure 7-7 (a) Schematic symbol for a triode. (b) The effect of the grid on the current flow. Small voltage changes on the grid result in large changes in the current flowing through the tube.

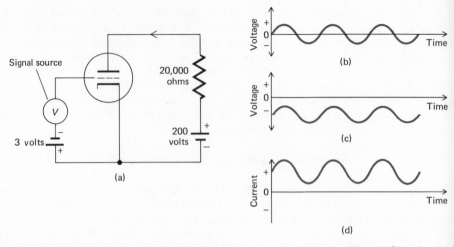

Figure 7-8   A simple triode amplifier. (a) Circuit diagram of a triode amplifier. (b) Signal voltag
(c) The total voltage on the grid of the triode. (d) The current through the triode.

tant. Power amplification is obtained in a triode circuit because, in addition to th
voltage, the current is also amplified. Let us consider a numerical example. Th
output voltage is 20 volts and the signal current is 1 milliampere (1 milliampe
is the change in the tube current produced by the 1-volt change on the grid). Th
output power (P_{out}) is therefore

$$P_{out} = V_{out} \times I_{out} = 20 \times 1 \times 10^{-3} = 20 \times 10^{-3} \text{ watt}$$

The input voltage is 1 volt and the input current is typically 1 microampere
microampere is 1-millionth ampere $= 10^{-6}$ ampere). Therefore the input pow
(P_{in}) is

$$P_{in} = V_{in} \times I_{in} = 1 \times 10^{-6} = 10^{-6} \text{ watt}$$

The power amplification is the ratio of the output to the input power; thus

$$\text{power amplification} = \frac{P_{out}}{P_{in}} = \frac{20 \times 10^{-3}}{10^{-6}} = 20,000$$

The additional power in the amplified signal is provided by the power supply th
biases the anode.

For simplicity's sake we have used sinusoidal signals to illustrate our discussio
of amplification. In practice the signals can be much more complex, but the bas
amplification process is independent of signal shape. Voltage amplification by
factor of 20 is typical of most triode amplifiers. Higher signal amplification can l
obtained by applying the output of one amplifier to the input of another.

Shortly after the invention of the triode, more electrodes were added to th
structure, producing the tetrode and the pentode. The operation of these mul
electrode tubes is basically the same as that of the triode, but they can provide
higher power and voltage amplification.

An amplifier can be considered as a device that multiplies the input by the a
plifying constant A (Figure 7-9a).

$$V_{out} = AV_{in} \tag{7-}$$

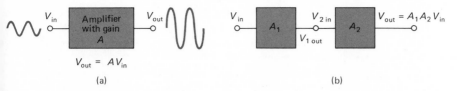

$V_{out} = AV_{in}$

(a)

(b)

gure 7-9 Amplification, (a) Block diagram of an amplifier, $V_{out} = AV_{in}$. (b) A signal amplified by two amplifiers, $V_{out} = A_1A_2V_{in}$.

For example, if a 2-volt signal is applied to the input of an amplifier that amplifies by a factor of 15 ($A = 15$), the output voltage is 30 volts. To obtain the amplification factor of an amplifier we can measure the output and input voltages and divide one by the other; that is,

$$A = \frac{V_{out}}{V_{in}}$$

When two amplifiers are connected together, the total amplification of the system is the product of the individual amplifications (Figure 7-9b); that is, if the output of an amplifier with amplification A_1 is applied to an amplifier with amplification A_2, the total amplification A_T of the signal is $A_T = A_1A_2$. This relationship is derived in the appendix to this chapter. In this way it is possible to obtain very high amplification. For example, if A_1 is 15 and A_2 is 20, the amplification of the combined system is 300.

In the few years following the invention of the triode its operation was greatly improved through the work of Hall, Langmuir, DeForest, Armstrong, and many others. The work of Armstrong is especially notable. During the period of 1911 to 1914 he explained the theory of the triode and fully explored its potential. He also discovered the principle of feedback which greatly increased the versatility of amplifiers.

EEDBACK AMPLIFIERS AND OSCILLATORS

An ordinary amplifier is converted into a feedback amplifier simply by applying part of the amplified signal back into the grid of the amplifier.

A simple feedback amplifier is shown in Figure 7-10. The signal applied directly into the amplifier is the sum of the externally applied signal V_{in} and part of the amplified signal βV_{out}. The symbol β (Greek beta) is called the *feedback* factor and is equal to the fraction of the voltage that is being fed back to the input. We show in Appendix 7-2 that the amplification of a feedback amplifier is

$$A_f = \frac{A}{1 - A\beta}$$

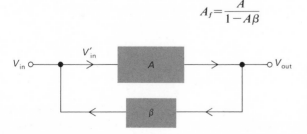

Figure 7-10 A feedback amplifier with feedback factor β. The total amplification of the system V_{out}/V_{in} is $A/(1 - A\beta)$.

where A is the amplification without feedback. One use of feedback is to increase the overall amplification of the system. For example, if the gain of the basic amplifier without feedback is 20 and the feedback factor is 0.04, then the amplification of the feedback system as a whole is

$$\frac{20}{1 - 0.8} = 100$$

We note that if $A\beta$ is one, then the denominator in the expression for the overall gain becomes zero, and therefore the total amplification becomes infinite. This implies that a finite input voltage produces an infinite output voltage. This, however cannot happen. There are limitations to the operation of all the circuit components in the amplifier. The power supply, for example, can supply only a limited amount of power and voltage. Therefore the output voltage reaches some maximum value determined by the capabilities of the amplifier. But by making $A\beta$ equal to one the amplifier can be converted into an oscillator, which is a device that produces sinusoidal output signal at a specific frequency. The circuit diagram of a feedback oscillator is shown in Figure 7-11. The differences between this circuit and the ordinary triode amplifier are immediately evident. The anode resistor is replaced by resonant circuit and part of output is fed back into the grid by a transformer. When a signal at the frequency of the resonant circuit is applied to the grid, the signal is amplified and a voltage appears across the resonant circuit. If the circuit is adjusted so that $A\beta$ is one, then the voltage fed back to the grid by the transformer causes the output to grow until it is limited by saturation. The output is a steady sinusoidal voltage with the frequency determined by the resonance circuit which $1/(2\pi\sqrt{LC})$. A circuit that produces a steady sinusoidal voltage at a specific frequency is called an *oscillator*.

We may next ask how this process is initiated. More exactly, what is the source of the voltage at the resonance frequency that starts the oscillation process? The source of this voltage is noise. In all electric circuits there are random fluctuations in current and voltage that are called *noise*. For example, the emission from the cathode of a vacuum tube is not completely uniform. It fluctuates around the average value by a small amount and therefore produces noise in the circuit. These random voltage and current variations are present at all frequencies and are sufficiently large to start the oscillation process.

Although there are many other types of oscillators besides the one we have described, they all work on the basic principle of feedback. Oscillators are widely used in communication systems for generating radio signals. Their usefulness will become evident as we discuss the radio in greater detail.

The principle of feedback has a much broader application in technology than we have implied in this discussion. Feedback is used in all self-regulating systems

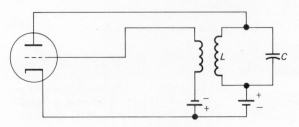

Figure 7-11 Triode oscillator. The frequency of oscillation is $1/2\pi\sqrt{LC}$.

It is the basis of automation in all biological, electrical, and mechanical systems. A thorough description of feedback and its applications is found in the references.

OPERATION OF A RADIO SYSTEM

We have now gathered enough technical background to describe the operation of the modern radio. Although many of the sophistications that are found in modern radio are omitted from our discussion, the main ideas are conveyed. The simplest way to describe the modern radio system is in terms of the block diagram shown in Figure 7-12. Each block in the diagram represents a necessary function in broadcasting. Some of these have already been discussed; others are described as we go along.

We have already described the basic principles of broadcasting. To recapitulate, at the transmitting end the voice information is superimposed onto an electromagnetic signal which is radiated into space. The electromagnetic wave is detected in the receiving part of the system. The voice information is extracted from the electromagnetic wave and is converted into sound. We now describe in detail how this is done. The microphone converts the acoustical sound waves into an electric signal which is called the *audio signal*. The audio signal is then amplified, and its information content is imposed on a high frequency radio signal which is radiated into space. Why do we not broadcast the audio signal directly? It is, after all, an electric signal and therefore could be radiated into space. This, in fact, could be done; but there are several reasons for not doing it. As mentioned earlier, the longer the wavelength of radiation, the larger is the antenna required to launch efficiently the radiation into space. The frequency of the audio signal is, of course, the same as the frequency of the sound wave. This frequency is low — between 20 and a few thousand cycles per second. In order to efficiently launch electromagnetic radiation in this frequency range, the transmitting antenna would have to be inordinately large. Furthermore, if broadcasting were done directly at audio frequencies, all broadcasters would be forced to transmit within the same frequency range, which would result in intolerable confusion. For these reasons broadcasting has to be done at frequencies considerably higher than the audio range. Thus the information contained in the audio signal has to be superimposed onto a high-frequency radio signal, which is then radiated into space. The high-frequency radio signal is called the *carrier* and the process by means of which the audio information is placed on the carrier is called *modulation*. The frequency of the carrier is typically above 500,000 cycles per second. In order to avoid interferences each station in a given broadcasting region uses a somewhat different carrier frequency.

We have already discussed one very simple modulation technique, the transmission of the Morse code. Here the carrier is simply turned on and off, transmitting pulses that convey the alphabet. The information is contained in the duration of the radio frequency pulse. A more sophisticated modulation technique has to be used if the full audio information is to be implanted onto the carrier.

The carrier frequency alone contains no information. In order to put the information contained in the audio wave onto the carrier, some feature of the carrier has to be varied in accord with the audio signal. Because there are two parameters that describe an electromagnetic wave, the amplitude and the frequency, we immediately have two ways of implanting the information onto the carrier. We can vary either the amplitude or the frequency of the carrier in accord with the audio signal. If the audio information is contained in the amplitude of the carrier, the modulation is

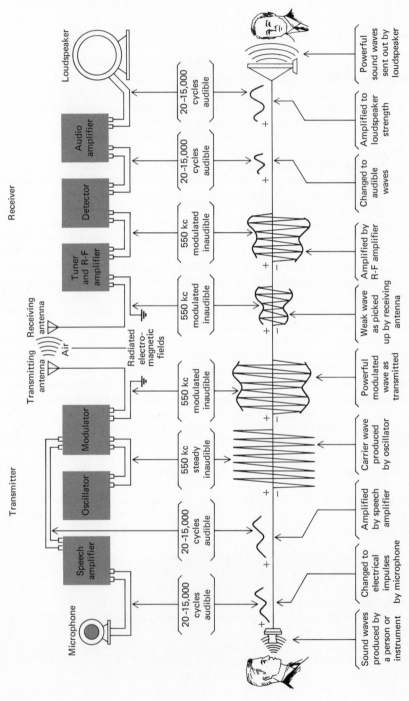

Figure 7-12 Radio-broadcasting operation from microphone to loudspeaker.

called *amplitude modulation* (AM). If the information is carried in the frequency of the carrier, the modulation is called *frequency modulation* (FM). At present both types of modulation are used, but because amplitude modulation is both technically and conceptually simpler, it is described first.

AMPLITUDE MODULATION (AM)

Returning to our block diagram (Figure 7-12), we can see that the carrier is generated by an oscillator which may be of the type we have discussed. In the modulator, the amplitude of the carrier signal is varied in accord with the audio signal to produce the modulated carrier (Figure 7-13). The amplitude of the modulated carrier has been outlined to point out that it follows the shape of the audio signal. Modulation can be accomplished in many ways, one of which is shown in Figure 7-14. In this modulating circuit both the carrier and the audio signal are applied to the grid of an amplifier. If the grid bias voltage V_c is properly adjusted, the amplifier output is the amplitude-modulated carrier. As shown in the block diagram the modulated signal is then radiated into space by the antenna. At the receiving station the electromagnetic radiation causes current to flow in the antenna.

Our space is filled with electromagnetic radiation at various carrier frequencies emitted by different broadcasting stations. Currents produced by all these carrier frequencies flow in the antenna and they have to be sorted out in some way. It is the tuner that picks out only that carrier frequency in which the listener is interested. In its simplest form the tuner consists of a capacitor-inductor resonant circuit tuned to the carrier frequency of interest. Tuning is usually done by changing the capacitance of a variable capacitor. Only the carrier wave at the tuned frequency produces appreciable current in the resonant circuit. The amplitude-modulated carrier is then amplified and applied to the detector which extracts the audio signal. The detector consists of a diode operating in a way that we have discussed before. The diode rectifies the carrier current and the capacitor smoothes the pulses so that the

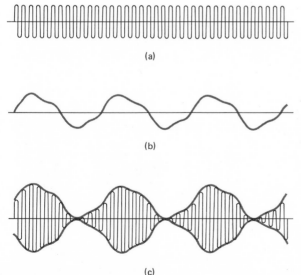

(a)

(b)

(c)

Figure 7-13 Carrier wave, voice signal, and modulated carrier wave. (a) Carrier wave which has constant amplitude and frequency. (b) Voice or audio signal which modulates the amplitude of the carrier. (c) Modulated carrier wave in which the amplitude of the carrier varies in accord with the audio signal. The envelope joining the peak of the carrier signal is drawn to indicate the variations in the carrier amplitude.

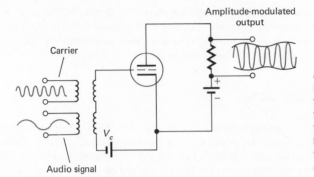

Figure 7-14 An amplitude modulator. The carrier and the audio signal are applied to the grid of a triode. If the triode is appropriately biased, the output will be the amplitude-modulated carrier.

voltage that appears at the output has the shape of the original audio signal. The audio signal is then amplified and applied to a speaker, where it is converted into an acoustical signal. The whole process is illustrated in Figure 7-15.

The development of radio was the work of hundreds of people; still there are a few men whose accomplishments in this field were outstanding. Among these is Lee DeForest (1873-1958). He is undisputedly recognized as the man who invented the triode which, as we have seen, is the key to amplification. DeForest himself made claim for just about everything that has been done in radio technology. Many of his claims, however, have been widely challenged, and through most of his professional life he was embroiled in patent litigations. In 1950 DeForest wrote his autobiography which he modestly titled *Father of Radio*. It is a literate, delightful book filled with a large amount of self-praise that to his contemporaries must have been jarring. From his diary and writings one recognizes a tremendous drive to be successful. He centered his efforts on inventing something, anything that would

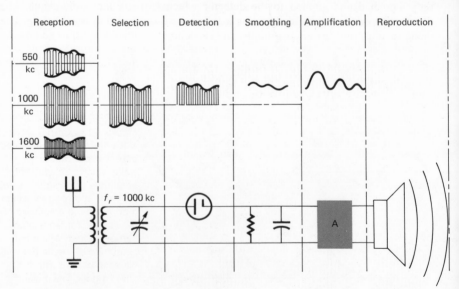

Figure 7-15 Diagram of diode detector receiving circuit and the type of current flowing through each of its parts.

gain him fame and fortune. He tried to invent a better subway system, steam engine, bicycle, telephone relay, but his initial inventive efforts were not successful. He attended Yale University both as an undergraduate and graduate student and in 1899 he received a doctorate degree. A year later he began working as an independent inventor. He was granted his first patent in 1902. Eventually he had more than 200 patents to his name. He was a feverish worker, pursuing many ideas simultaneously. It is not clear how much of his work was original. From reading the description of his work it appears that in many cases he did not have a clear understanding of what he was doing. It is obvious that he had a great and wonderful intuition, but many of his explanations and ideas were wrong. His work touched on most developments in electronics, but he also produced sound motion pictures and built loudspeakers. In the 1930s he experimented with television, and in 1948 he patented a color television. On and off he had collaborators, but these unions never lasted more than a couple of years. He had a deep and sound concern about the future of communication. In 1946 he wrote an open letter to the Convention of the National Association of Broadcasters:

A Father Mourns His Child

In the Palmer House is assembled the Convention of the National Association of Broadcasters. There many words are being spoken on behalf of a great industry which thrives chiefly on spoken words.

One wonders if our simian ancestors had any conception that in ages to come such monkey chatter as they originated would be transformed into the essentials of livelihood. Of such are the mysteries of evolution. Today fabulous sums are paid for talk. Speech, not silence, has proved golden; and the dispensers of such merchandise to the millions are here foregathered to plan for more speech, for more money.

Had I, who originated the idea and the means for broadcasting, been invited to their council, I should say to them: "What have you gentlemen done with my child? He was conceived as a potent instrumentality for culture, fine music, the uplifting of America's mass intelligence. You have debased this child, you have sent him out in the streets in rags of ragtime, tatters of jive and boogie woogie, to collect money from all and sundry, for hubba hubba and audio jitterbug. You have made of him a laughing stock to the intelligence, surely a stench in the nostrils of the gods of the ionosphere; you have cut time into tiny parcels called spots (more rightly stains), wherewith the occasional fine program is periodically smeared with impudent insistence to buy or try.

Murder mysteries rule the waves by night and children are rendered psychopathic by your bedtime stories. This child of mine, now thirty years in age, has been resolutely kept to the average intelligence of thirteen years, as though you and your sponsors believe the majority of listeners have only moron minds. Nay, the curse of your commercials has grown consistently more cursed, year by year.

But his British brother has had a different upbringing. Under government sponsorship, radio appeals there to the higher intelligence, realizing its fine mission to elevate, and not degrade. This is anathema to America's broadcasters — vastly enriched by their freedom from all restraint. We prefer to pay the colossal bill in gold, and in the debased coinage of the anesthetized intellect. We might learn much from England."

Yet, withal, I am still proud of my child. Here and there from every station come each day some brief flashes worth the hearing, some symphony, some in-

telligent debate, some playlet worth the wattage. The average mind is slowl
broadening, and despite all the debasement of most of radio's offerings, ou
music tastes are slowly advancing.

Some day the program director will attain the intelligent skill of the engineer
who erected his towers and built the marvel which he now so ineptly uses.[1]

DeForest also wrote, "Let us fervently hope that FM will prove to be also
new tongue to give the world programs of the highest where its older sister has ofte
so lamentably failed. And if not FM, television will soon free millions of listener
from the blighting curse which has so long debased AM broadcasting."

Unfortunately, DeForest in this prediction was mostly wrong.

FREQUENCY MODULATION (FM)

Although amplitude modulation is the simplest method of modulation, it has
major fault. Most natural and man-made electrical disturbances can impress them
selves on the carrier wave and produce amplitude variations on it. At the receivin
station these unwanted amplitude variations are detected and produce noise. Fo
example, a flash of lightning produces a pulse of electromagnetic radiation tha
passes through the receiver as a crackle. Electric motors, generators, and trans
formers induce similar disturbances. As we have previously mentioned there i
another way of modulating a carrier wave and this is by varying its frequency i
accord with the audio signal. Frequency modulation removes the system from th
domain of these noise sources. In frequency modulation the frequency of the carrie
is varied by an amount proportional to the amplitude of the sound. The rate at whic
this variation takes place is determined by the frequency of the sound wave. Thi
is illustrated in Figure 7-16. A large amplitude audio signal produces a correspond
ingly large frequency change in the carrier. The amplitude of the carrier is no
changed in the process of modulation. Since the amplitude of the carrier contain
no information, any amplitude variations produced by noise can be eliminated b
clipping them with a suitable device (Figure 7-17).

Although modern *frequency modulators* are complex, their basic principles o
operation are easily understood. Let us consider the oscillator circuit shown in Fig
ure 7-18. This circuit will oscillate at the resonance frequency which is $1/(2\pi\sqrt{LC})$
If the circuit capacitance is varied, the frequency of oscillation will vary accord

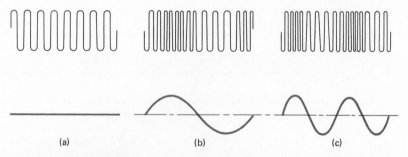

(a)   (b)   (c)

Figure 7-16 Frequency modulation. The frequency of the carrier wave (a) is changed by th
modulating signal (b) and (c).

[1]Lee DeForest, *Father of Radio,* Follett Publishing Company, Chicago, 1950.

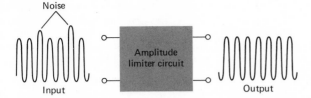

Figure 7-17 Limiter circuit. Amplitude variations due to noise are eliminated.

ingly. It is possible to construct a resonant circuit so that its capacitance is changed by the amplitude of the audio signal. Thus an increase in the amplitude of the audio signal decreases the capacitance and therefore increases the oscillation frequency, whereas a decrease in the amplitude of the audio signal has the opposite effect. In this way the amplitude variations of the audio signal are transformed into frequency variations of the carrier.

As in amplitude modulation, here also the frequency-modulated carrier is launched into space and is received by an antenna. The carrier now has to be demodulated or, in other words, the frequency changes in the carrier have to be converted in an amplitude-varying audio signal that can drive a speaker. In principle, at least, the frequency demodulation process is again simple. Let us consider the tuned amplifier shown in Figure 7-19a. The amplification of this circuit is largest when the frequency of the input signal is at the circuit resonance frequency (f_o). The amplification decreases at both sides of the resonance frequency as shown in Figure 7-19b. This circuit can convert the frequency-modulated carrier into an amplitude-modulated carrier from which the audio signal is extracted with a conventional diode detector. Conversion from frequency to amplitude modulation is done in the following way. The resonance frequency of the circuit is adjusted to be higher than the frequency of the unmodulated carrier (f_c). Operation in this region of the amplification curve results in a frequency-dependent amplification. Thus when the frequency of the modulated carrier increases, the amplification increases, which results, of course, in a larger output voltage. Similarly, a decrease in the carrier frequency results in a smaller output voltage. In this way the frequency variations of the carrier are converted into amplitude variations (Figure 7-19c). The frequency variations of the original signal are retained in the output. This, however, is of no consequence because the diode detector to which this signal is applied is not sensitive to the frequency variations of the carrier. The audio signal is therefore detected and converted to sound as in amplitude modulation.

Whereas amplitude-modulated broadcasting was the work of many people, the frequency-modulation (FM) radio system was entirely the invention of Edwin H. Armstrong. I first came into contact with Armstrong's work in 1958 when I was an undergraduate at Columbia University. I was working as a technician in a small electronics laboratory that was in the basement of the psychology building. Every

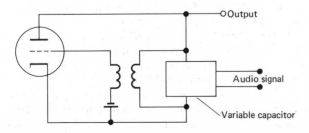

Figure 7-18 A frequency modulator. The capacitance of the variable capacitor is varied by the amplitude of audio signal. The oscillation frequency of the circuit is therefore proportional to amplitude of the audio signal.

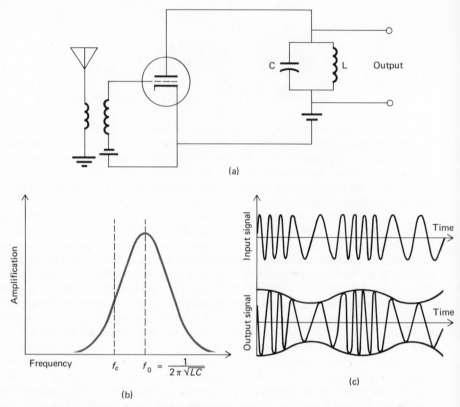

Figure 7-19 A frequency demodulator. The frequency of a tuned amplifier (a) is adjusted to be higher than the center frequency of the modulated carrier. Due to the frequency dependent amplification (b) the frequency variations are converted into amplitude variations (c).

afternoon about 3:00 P.M. an elderly gentleman arrived and carefully opened the big oak door that was adjacent to our laboratory. He went in, locked the door behind him, and then emerged again about three hours later. I was curious about this activity, but the technician I worked for did not know what was behind the oak door. After about a month, I asked the man what he was doing there. He invited me into his room. Inside there was a most beautiful laboratory with large amounts of carefully hand-constructed equipment, radio receivers, hundreds of tubes, an acoustically insulated chamber, and various antennas. This had been Armstrong's laboratory. He had committed suicide in 1954 (by jumping out of the window of his thirteenth floor New York apartment) and now this man, who had been his secretary, was carefully cataloging all of the Armstrong memorabilia.

Armstrong was born in New York City in 1890. While still a boy he was keenly interested in science and read widely in it. In 1909 he enrolled in electrical engineering at Columbia University. During his junior year he invented the principle of feedback. He filed a patent for his invention in 1913, but at this time no one seemed very interested in it. When the war broke out, he joined the Army and became a signal corps major in France. During this time he invented the superheterodyne receiver, which was a great improvement on the standard radio receiver described earlier. After the war he sold both patents to Westinghouse for $530,000. In 1922 Armstrong sold another patent to the Radio Corporation of America (RCA) for $200,000 and 80,000 RCA shares which he sold for $100 each just before the 1929 stock crash. At this time and to the end of his life Armstrong was a Professor of Electrical Engineering at Columbia University. About 1925 he began working on frequency modulation. He completed the frequency-modulated system in 1933 and demonstrated it to the Institute of Radio Engineers in 1935. Armstrong had a number of frequency-modulation radio sets behind him as he described the system operation. When he finished, he pointed out to his audience that while he was talking the speakers were on full volume but without an incoming signal. Yet not a sound came out of them.

This was quite an impressive demonstration since an amplitude-modulation set cannot be turned on without hearing static and hum from it. He then phoned his technician at the transmitter station and proceeded to demonstrate high-fidelity sound reproduction. The sound of a wrinkled paper, the striking of a match, and some music were all transmitted with great clarity. The radio industry, however, paid little attention to frequency modulation. They had too much invested in amplitude modulation and were doing too well financially to bother with a new system, even if it was superior. The Federal Communications Commission was not much more sympathetic. The Commission stalled in giving Armstrong a permit to build an experimental station. Armstrong threatened to build the station in some other country and finally in 1937 he was given a permit for construction. He built his station at a personal cost of $300,000 in Alpine, New Jersey. During the next few years he gave demonstrations of the superior quality of frequency modulation, but only a few people, including those to whom he had given receivers, were able to listen to his broadcasts.

Almost from the beginning of his professional career Armstrong was embroiled in patent fights. A year after Armstrong applied for his feedback patent, DeForest challenged it. The radio industry first backed Armstrong and then switched its backing to DeForest. In 1924 DeForest was granted the patent. Armstrong was very upset. He decided to return the award that the Institute of Radio Engineers had given him for his inventions in 1917. The organization refused to accept it back. In 1941 and 1942 a technical jury again reexamined the invention claims and decided to award Armstrong the Franklin Medal for his inventions.

Just before World War II FM radio began to receive more attention. Stations began to multiply and by 1949 there were 600 frequency-modulation stations. The field was now big business. Armstrong refused to sell his patent for the system. He decided to rent only the rights, thereby hoping to control the system and prevent it from becoming like amplitude-modulation broadcasting. But he failed in this. Companies pirated his patents and manufactured frequency-modulation sets without his consent. Although some frequency-modulation stations maintained high-quality broadcasting, many of them deteriorated and a good number of them simply duplicated amplitude-modulation programs. Armstrong kept his Alpine station operating at his own expense until his death.

Toward the end of his life Armstrong was again involved in a long patent battle over the frequency-modulation rights. People who knew him remember the great sense of failure Armstrong had during this period. Had he lived he may have been somewhat consoled by the present state of FM. Certainly much of FM broadcasting is not of the quality Armstrong hoped for. Still in most localities there is at least one FM station which broadcasts the type programs Armstrong would have liked.

APPENDIX 7-1

Combined Gain of Two Amplifiers (Refer to Figure 7-9)

The combined amplification

$$A_{\text{total}} = \frac{V_{2\text{ out}}}{V_{\text{in}}}$$

where

$$V_{2\text{ out}} = A_2 V_{2\text{ in}}$$

But

$$V_{2\text{ in}} = V_{1\text{ out}} = A_1 V_{1\text{ in}}$$

Therefore

$$V_{2\text{ out}} = A_2 A_1 V_{1\text{ in}}$$

and

$$A_{\text{total}} = \frac{A_2 A_1 V_{1\text{ in}}}{V_{1\text{ in}}} = A_2 A_1.$$

APPENDIX 7-2

Amplification of a System with Feedback (Refer to Figure 7-10)

$$A_{\text{feedback}} = \frac{V_{\text{out}}}{V_{\text{in}}} \tag{1}$$

Here V_{in} is the input signal voltage from some external source. The actual input voltage V'_{in} into the amplifier is the sum of the external signal and the signal that is fed back from the output.

$$V'_{\text{in}} = V_{\text{in}} + \beta \times V_{\text{out}} \tag{2}$$

But

$$V_{\text{out}} = A V'_{\text{in}}$$

Through substitution into Equation (2) we get

$$V'_{\text{in}} = V_{\text{in}} + A\beta V'_{\text{in}} \qquad \text{or} \qquad V_{\text{in}} = V'_{\text{in}}(1 - A\beta)$$

Therefore the total gain with feedback is

$$A_{\text{feedback}} = \frac{V_{\text{out}}}{V'_{\text{in}}(1 - A\beta)} = \frac{A}{1 - A\beta}$$

8 Reproduction of Sound

In Chapter 7 we were concerned mostly with the technology of the interconversion of electric signals and sound. In this chapter we discuss the factors that determine the quality of the sound reproduction. This discussion also applies to sound reproduction from tapes and records, subjects we have not treated so far. We therefore begin with a brief description of phonographs and tape recorders.

PHONOGRAPH AND TAPE RECORDING

A basic block diagram of a phonograph or tape-recording system is shown in Figure 8-1. The microphone converts the sound into the electric audio signal that is amplified to a level necessary for storing the sound information on a record or tape. On replay the stored sound information is converted into an audio signal which, after amplification, drives a speaker or earphones.

The first recording equipment was built in Edison's laboratory in 1880. By an arrangement of levers the pressure variations of the sound wave were made to drive a sharp needle into a wax cylinder. As the cylinder was rotated at a constant rate, the needle was guided from one end of the cylinder to the other. Since the force applied to the needle was proportional to the pressure variations of the sound, the groove cut by the needle contained "hills and valleys" with heights proportional to the sound amplitude. On playing back, the cylinder was rotated at the same speed as the recording, and a needle, often made of wood, bounced up and down as it traveled along the groove. The needle was connected to a horn that vibrated in

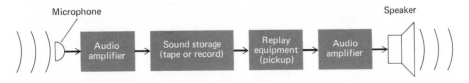

Figure 8-1 A block diagram of a sound-reproduction system.

synchrony and reproduced the sound. The power to operate the needle and the horn came entirely from the initial sound. As can be imagined, the quality of reproduction obtained from these early phonographs was poor.

With minor modifications the basic principle of phonograph recording has remained unchanged since Edison's time, but the use of electronic amplification and sophisticated recording techniques has made the phonograph an excellent sound-reproducing instrument. The amplified audio signal produces a magnetic field which drives a sharp cutter into a plastic disk (Figure 8-2). This conversion of the electric audio signal into the mechanical motion of the cutter is based on the principle we described in our discussion of speakers. The cutter is guided in a spiral path from the outer rim to the center of the disk. In modern recording the sound vibrations produce a side-to-side motion of the cutter, which is a departure from the up-and-down motion used in the original cylinder recordings. The wavy pattern of the groove is a replica of the loudness and frequency of the driving sound. In the commercial manufacture of records master dies are made and these are used to mold plastic records.

Sound reproduction from a record is essentially the reverse of the recording process. As the disk rotates at a constant speed the phonograph needle moves along the wavy pattern of the groove. This motion is transmitted to a magnet which induces an audio current in a coil. The amplified audio current drives the speaker. Instead of magnetic induction, *piezoelectric crystals* are often used to produce the audio current. Piezoelectric crystals produce a voltage proportional to the pressure applied on their surfaces. Here the pressure is applied to the crystal by the moving needle and therefore the piezoelectric voltage is a replica of the groove pattern.

Experiments with sound reproduction from *magnetic tape* began in about 1898 when Valdemar Poulsen, a Danish scientist, discovered that varying audio currents produced by a microphone can induce magnetization in a steel wire. The degree of

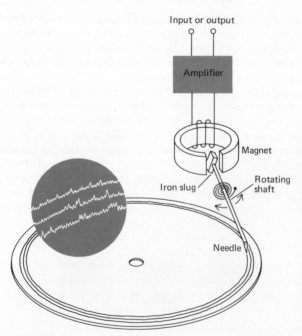

Figure 8-2 Phonograph recording and reproduction. In recording, the amplified audio signal produces a magnetic field which moves the needle. The wavy pattern of the groove produced by the needle contains the sound information. During playback the needle moves in the groove. The motion of the needle induces a current in the coil wound around the magnet. After amplification this audio signal drives the speaker.

magnetization is proportional to the current and therefore to the sound that produces it. When the magnetized wire is passed through the poles of an electromagnet, the changing magnetization induces an electric current which is a reproduction of the original current and can be used to drive a microphone (Figure 8-3). The original device was inadequate. The obtainable magnetization was too small to operate a useful speaker. With the advent of amplification the idea was revived in Germany and in the 1920s the first tape recorders were being built there. Magnetized wires were difficult to work with and expensive to manufacture. It was soon discovered that magnetic recording could also be made on a plastic tape coated with magnetic powder. The principle is the same; the audio current causes an alignment of the magnetic particles on the tape. By the late 1930s tape recorders of the type we have today were in use.

Sound storage on tapes is in many ways superior to storage on phonograph records. As we shall soon explain, the range of sound volume recordable with tapes is larger than with records, and wear due to replay is significantly lower than on records. The magnetization of a tape deteriorates with time, however, especially under unfavorable temperature and humidity conditions. It has also been found that as the tape is stored, magnetization from one layer tends to imprint on the adjacent layer. This often results in an effect called *pre-echo*.

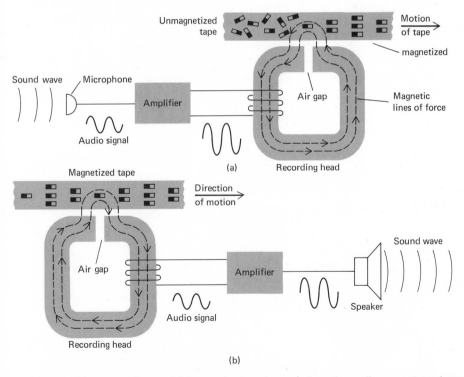

Figure 8-3 (a) Tape recording and (b) reproduction. In recording, the audio current produces a magnetic field which produces a corresponding magnetization on the tape. In reproduction, the moving magnetized tape induces a current in the coil wound around the magnet of the recording head. This current is proportional to the tape magnetization.

QUALITY OF SOUND REPRODUCTION

We have already mentioned that the wave pattern of sound is not usually of the simple sinusoidal form that we used to illustrate our discussions. In most cases the wave patterns produced by instruments and voices are highly complex (see Figure 4-5). As we have mentioned before it is this detailed pattern of the sound that determines its nature. The performance of a sound-reproducing system is judged by how well it preserves the wave pattern of the sound from input to output

It would be impossible to evaluate the performance of a system if we had to test its response to all possible wave shapes. Fortunately the problem is not that complicated. About 150 years ago, Fourier, a French mathematician, showed that all complex wave shapes can be analyzed into simple sinusoidal waves of different frequencies. In other words, any complex wave pattern can be constructed by adding together a sufficient number of sinusoidal waves of appropriate frequencies and amplitudes. Such an analysis of a wave shape is shown in Figure 8-4. The lowest frequency in the wave form is called the *fundamental* and the higher frequencies are called *harmonics*. For a given note played by the various instruments shown in Figure 4-5 the fundamental frequency is the same, but the harmonic content of the wave is different for each instrument. It is the harmonic content of the sound that differentiates one sound source from another. Therefore, in order to evaluate the response of the sound system to a given wave pattern, it is sufficient to determine the system response to the individual sinusoidal components.

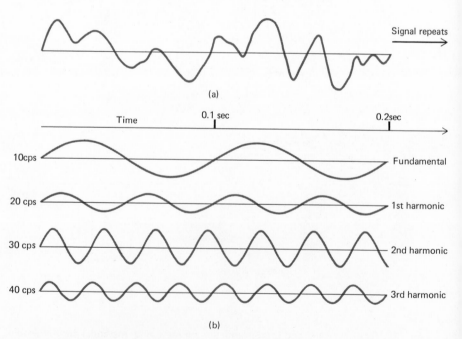

Figure 8-4 The analyses of a complex wave shape (a) into its sine components (b). The point by point addition of the fundamental frequency sine wave and the harmonic frequency sine waves yields the wave shape shown in (a).

If the fundamental and all the harmonics are faithfully reproduced, we can be certain that the wave shape of the output is identical to the input. Therefore, in order to determine what the reproduction system does to the wave as a whole, it is sufficient to know how the system responds to sinusoidal waves of different frequencies. For this purpose *frequency response* curves are constructed for the various components of the sound-reproduction system. These plots show the relative size of the output signals for the same size sinusoidal input signals over the entire frequency range. Figure 8-5a shows an ideal frequency-response curve. The straight line signifies that the system responds equally well to sinusoidal waves at all frequencies. A system with an ideal frequency response will reproduce the input exactly. Unfortunately no part of the sound system has an ideal frequency response. For example, for a given driving force the microphone may not vibrate as efficiently at high frequencies as at low frequencies. A good, realistic frequency-response curve for a microphone is shown in Figure 8-5b. As we can see, the curve droops both at high and low frequencies. This means that in converting the sound

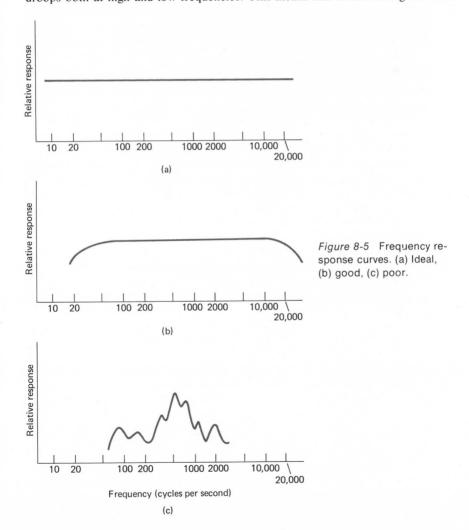

Figure 8-5 Frequency response curves. (a) Ideal, (b) good, (c) poor.

into an electric signal both the low- and the high-frequency components of the signal will be somewhat suppressed. If the microphone frequency response is poor as shown in Figure 8-5c, the sound will be badly distorted (Figure 8-6). Fortunately the human ear does not respond to all the features of the sound. A system can cause considerable alteration of the sound wave without a noticeable effect on the audio sense. Therefore in our discussion of sound-reproduction quality we must also consider the audio response of the ear.

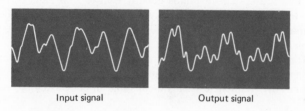

Input signal   Output signal

Figure 8-6 Distortion of the input signal by a microphone with a poor frequency response.

The frequency response of the human ear is limited. An average person can hear signals in the frequency range from about 30 to 20,000 cycles per second. Although there are variations in the audio frequency response of individuals, the response does not extend much beyond these limits. The frequency response of the sound system outside these limits is of no importance to the listener.

Although there is still some dispute about this, it appears that the ear does not distinguish the relative phases of the harmonic components in the sound signal. For example, the signal shown in Figure 8-7a is of a completely different shape from the signal in Figure 8-7b. However, the amplitudes of the sinusoidal frequency components are identical for the two signals. The difference between the two signals is the relative phase or displacement of the harmonic components. Since the ear responds only to the amplitude of the component sinusoidal waves, it cannot distinguish between the two signals. Therefore we do not have to worry about signal distortion caused by phase shifting in the sound-reproduction system. On the other hand, within its frequency response the ear is sensitive to frequency distortions. The suppression of low-frequency components makes for a shrill, tinny sound. The elimination of high-frequency components causes the sound to be muffled. Frequency distortion in the middle range reduces the clarity of the sound.

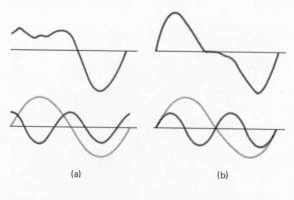

(a)   (b)

Figure 8-7 Since the frequency components of signals (a) and (b) are identical, the ear does not distinguish between them.

SPEAKERS AND MICROPHONES

The frequency response of a modern sound-reproducing system can be made excellent. Still the system must be designed properly if it is to reproduce faithfully the input sound. Typically, the microphone and the speaker are the components that must be designed with the greatest care. We have already described the operation of the induction microphone. There is, however, another type of microphone operating on electrostatic principles that is more commonly used today in high-quality reproduction systems. In the electrostatic microphone a charge is maintained on the plates of the capacitor. The pressure from the sound causes the charged plates to move together in synchrony with the sound variations. This motion of the charged plates produces a current corresponding to the sound variations. The principle of the electrostatic microphone was invented very early in the history of the electrical communication, but the device was useless without electrical amplification.

The most commonly used speakers are of the electromagnetic type described earlier. The properties of this type of speaker were studied in the 1920s by C.W. Rice and E.W. Kellogg. They found that large speakers reproduce low-frequency sounds well, whereas small ones were better for high-frequency reproduction. There have been some successful designs of speakers with relatively good performance over the whole sound frequency range, but in most good sound systems two or more speakers are used, each designed for a specific frequency range.

Recently electrostatic speakers have come into wider use. These work on the same principle as the capacitor microphone. The variation of charges produced by the audio signal causes a diaphragm to move and reproduce the sound. The electrostatic speakers have an excellent frequency response, but at present they are more expensive than the electromagnetic speakers.

SOURCE OF SOUND DISTORTION

A basic problem of high-fidelity sound reproduction is the distribution of sound from the speaker. Sound is emitted by both the front and back of the speaker. The sound is generated at the same time, but the sound from the back of the speaker must travel further to reach the listener. As shown in Figure 8-8 this can cause interference between the two sound patterns and result in a muffled effect. The

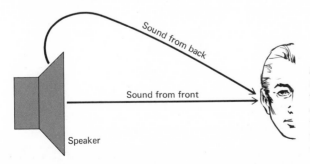

Sound from back

Sound from front

Speaker

Figure 8-8 Sound from the rear of the speaker reaches the ear after the sound originating from the front of the speaker. This interference may cause distortion of sound.

problem can be avoided by various baffling techniques which eliminate the sound from the back or alternately change the interference pattern at the listening position. In general the weakest links in a high-fidelity system are its speakers and it is therefore important to choose them with care.

An important source of bad sound reproduction is saturation. As the word implies some part of the reproduction system, for example, the amplifier, may not respond to the full amplitude of the input signal, thus causing the tops of the wave to be cut off (Figure 8-9). The flattening of the signal introduces additional frequency components which distort the sound. This distortion is dependent on the size of the input signal. Larger signals are proportionately more flattened and distorted than the smaller ones. It is interesting to note that systems with a wide frequency response are more susceptible to saturation distortion. Experiments conducted by Harry F. Olson in the late 1940s showed that with a system frequency response of 50 to 5000 cycles listeners accepted three times the distortion they could tolerate when the frequency response extended to 15,000 cycles per second. This is simply due to the fact that a narrower frequency response eliminates some of the unwanted frequency components introduced by saturation.

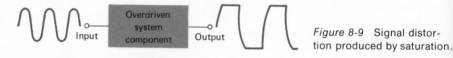

Figure 8-9 Signal distortion produced by saturation.

An important property of sound is its volume or loudness. We already know that sound moves objects such as the membrane of the microphone. We therefore conclude that sound can do work and carry power that is proportional to its loudness. In fact the loudness of sound is measured in units of watts per square centimeter (watts/cm²). At 1000 cycles per second the faintest sound the ear can respond to is about 10^{-16} watt/cm². This is the sound level of a normal voice at about 40 yards. The range of sound power that the ear responds to is enormous. The loudest sound the ear can tolerate is about 10^{-4} watt/cm², which is roughly the sound level produced by a jet aircraft at a close distance; that is, the loudest sound that the ear can tolerate is 10^{12} or a thousand-billion times more powerful than the softest sound it can detect. Fortunately the ear does not respond linearly to sound intensity. If the sound level is increased from 10^{-14} watt/cm², which is the level of a whisper, to 10^{-9} watt/cm², which is the sound level of a busy traffic-filled street, the ear detects an increase in loudness of a factor of about 4 rather than 100,000 which is the actual increase in sound power. The ear responds more or less logarithmically to the power increase.[1] For this reason it is customary to designate sound levels in logarithmic units. The threshold of hearing (10^{-16} watt/cm²) is taken

[1]Any number x can be expressed as $x = 10^y$. For example,

$$1 = 10^0$$
$$10 = 10^1$$
$$100 = 10 \times 10 = 10^2$$
$$1000 = 10 \times 10 \times 10 = 10^3$$
$$3 = 10^{0.477}$$
$$30 = 10 \times 3 = 10^1 \times 10^{0.477} = 10^{1.477}$$

In this representation y is called the logarithm (or log) of x. Thus the log of 10 is 1, the log of 1000 is 3, and the log of 30 is 1.477.

as the reference intensity level. The intensity of a sound P watts/cm^2 is then defined as

$$\text{sound intensity} = 10 \log \frac{P}{10^{-16}}$$

Intensity is expressed in units of *decibels* (abbreviated db). (Originally the intensity was defined as $\log P/10^{-16}$ in units of bels, named after Alexander Graham Bell. However, this unit proved to be too large and the decibel, which is one-tenth of a bel, was substituted for it.) The decibel value of sound intensity is simple to calculate. As an example, let us consider the sound intensity of a traffic-filled street ($P = 10^{-9}$ watt/cm^2):

$$\text{sound intensity} = 10 \log \frac{10^{-9}}{10^{-16}} = 10 \log 10^7 = 70 \text{ db}$$

The sound levels of various sources are shown in Table 8-1. As we can see, a quiet radio sounds twice as loud to our ear as a whisper and four times as loud as the rustle of leaves, even though the power ratios are enormously larger.

One of the biggest problems in sound reproduction is the range of volume to which the system can properly respond. The loudest reproducible sound is limited by the ability of the various components to operate without distortion. The lowest reproducible sound is determined by the inherent noise in the system. The reproducible volume range is called the *dynamic range*. A quantitative measure of the dynamic range is defined similarly to the sound intensity:

$$\text{dynamic range} = 10 \log \frac{P_{\text{loudest}}}{P_{\text{lowest}}}.$$

The limitations imposed by the obtainable dynamic range are specially severe for records. Records have a range of only 40 decibels. Although this is a reasonably wide volume range which extends from the loudness of a moderate conversation to the roar of an elevated train, the range is not nearly sufficient to accommodate the volume range of a full orchestra, which is about 80 decibels. The dynamic range of tapes is about 65 decibels, which is significantly better than the range of records but still not completely adequate.

A frequent and annoying source of sound distortion is noise. As we have mentioned before, noise is caused by spurious voltages added to the main signal. Noise can originate from external sources as well as from the components of the reproduction system. Much of the noise in radio transmission is produced by external

Table 8-1 Sound Levels due to Various Sources
(Representative Values)

Source of Sound	Sound Level (db)	Sound Level (watt/cm²)
Threshold of pain	120	10^{-4}
Riveter	90	10^{-7}
Busy street traffic	70	10^{-9}
Ordinary conversation	60	10^{-10}
Quiet automobile	50	10^{-11}
Quiet radio at home	40	10^{-12}
Average whisper	20	10^{-14}
Rustle of leaves	10	10^{-15}
Threshold of hearing	0	10^{-16}

electrical sources, such as lightning and electrical power equipment. For reasons we mentioned earlier, frequency-modulation broadcasting is much less susceptible to this type of noise than amplitude-modulation. Noise is frequently introduced by the tapes and records used in reproduction. The main sources of noise from records are warping, wear, electrostatic charges, and dust deposited on the disk. Reproduction from tapes is often distorted by pre-echo and residual tape magnetization.

SPATIAL PROPERTIES OF SOUND

Acoustics are of prime importance in a concert hall. The reverberations and echoes are an integral part of the performance, and a badly designed hall can ruin the performance of the best orchestra. Similarly, a perfect electronic sound system will produce bad sound if the room and the position of the speakers are not chosen properly. In listening to sound our ability to detect the location of the source depends on hearing with both ears. The ear closer to the source receives the sound earlier and also somewhat louder. The slight differences in the time of arrival and the intensity of the sound at the two ears are interpreted by the brain and produce the spatial image of sound. To fully perceive the sound spatially, each ear must hear something different. This stereophonic nature of hearing was recognized a long time ago and experiments in stereophonic systems began in the 1920s in England, Germany, the United States, and Denmark.

In the usual method of stereophonic recording the two microphones are separated by a distance that is determined by the type of performance being recorded (Figure 8-10). The two separate audio signals are imprinted in the record on the same groove. In the subsequent replay the needle follows the signal of both channels simultaneously, but the motions due to each audio signal are at right angles. The two signals are separated electronically, amplified, and transmitted to the respective speakers or

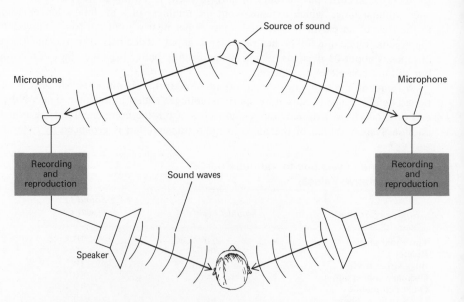

Figure 8-10 Principle of stereophonic sound recording and reproduction.

earphones. Recording with tapes is even more flexible since as many as four separate audio signals can be recorded simultaneously on the same tape. In stereophonic radio transmission two audio signals are transmitted and then electronically separated by the receiver. For the best stereophonic effect the two speakers in the listening area should be about as far from each other as the listener is from the speakers. The listening room should not be too vibrant, otherwise echoes will distort the sound.

Probably the goal of a good recording and its subsequent replay should be to reproduce the sound as it would be heard in a good auditorium. The definition of a "good" auditorium is highly subjective. The total sound is a result of a complex interplay of echoes and reverberations; the sound is even affected by the clothing of the audience. A recording completely capturing the concert hall sensation can be produced if a dummy is seated in a full concert hall and its ears are replaced by two microphones. The output of each microphone is recorded on a separate channel of a record or a tape. To reproduce the sound faithfully the listener has to use earphones, each of which is driven by the channel corresponding to the recording. Earphones must be used because if the sound is released by a loudspeaker, the new echoes and reverberations produced by the room alter the original sound. Although recordings of this sort have been made, they are clearly not suited for loudspeaker reproduction.

An audio technique that is coming into wider use is ambiophony. Speakers are placed in different portions of the room and the audio signals to them are delayed and controlled in such a way that the speakers in the rear of the room give an impression of reflection from a large concert hall. An increase in delay of the sound to the rear speakers produces an illusion of increased room size. Similar systems have been used to improve and tailor the acoustics of concert halls.

9 Television

DEVELOPMENT OF PICTURE TRANSMISSION

To those who have been reared in a technological environment television may seem no more mysterious than radio or any of the other devices that permeate our lives. But to people who first encounter technology, television is most perplexing. It is not unusual to hear a disembodied sound, but to see changing images without an obvious source is truly puzzling. Television is, in fact, a much more complicated device than radio even though the basic ideas are relatively simple.

A direct visual image is conveyed by light which is itself high-frequency electromagnetic radiation. However, the propagation properties of light and human optics are such that it is not possible for us to detect a detailed image from distances further than a few hundred yards. In order to transmit the visual image over long distances, the image information is converted into an electric signal which modulates a radio-frequency carrier. The modulated carrier is then radiated into space, picked up by the receiving antenna, and demodulated. The resulting signal is reconverted into a visual image. The process is therefore similar to radio broadcasting. The additional complexities arise in the conversion of the visual images into electric signals and vice versa.

It is interesting to trace the early attempts at transmitting images over long distances. Although they bear little resemblance to modern television, they point out some of the difficulties of image transmission. In 1873 it was discovered that the element selenium possesses a peculiar characteristic. Its resistance decreases when it is illuminated by light. (The reason for this effect is discussed in the next chapter.) When light shines on a selenium resistor that is connected as in the simple circuit shown in Figure 9-1, the current through the circuit increases because of the decrease in the resistance produced by the light. Since the decrease in the resistance is proportional to the light intensity, the current in the circuit is also proportional to it. An early suggestion for transmitting a picture using this effect was made in 1876 by G. R. Carey of Boston. He proposed to construct a board of many small selenium elements on which a picture is focused in much the same way as a camera focuses a scene on a film. The resistance of each selenium cell is then proportional

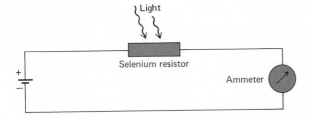

Light

Selenium resistor

Ammeter

+
−

Figure 9-1 The light decreases the selenium re-
sistance causing an increase
in the circuit current.

to the brightness of the image part that is shined on it. The image receiver consists of a board of small light bulbs positioned in an array identical to the selenium cells. Each light bulb is connected through a battery to the corresponding selenium cell. The current through each light bulb is thus proportional to the light intensity on the transmitting selenium cell. Since the brightness of each bulb is proportional to the current flowing through it, there is an element-by-element correspondence between the brightness of the transmitted and received images. Because the light bulbs are of finite size, the received image is granular; that is, the picture is made of elements with discrete brightness levels, whereas the brightness of the original image varies smoothly. This type image coarseness caused by the finite size of the transmitted image elements is characteristic of all present image transmission systems. By making the individual image elements small, however, the image can be made to appear tolerably smooth. It appears that some work was done on Carey's picture transmission device, but there are no reports of it ever being completed.

A more successful method of picture transmission, the facsimile, was perfected in the early 1920s and is still in use. In facsimile transmission the image is converted into an electric signal by the use of the photoelectric effect first observed by Hertz in 1887 during his experiments with electromagnetic radiation. Hertz observed that the production of sparks in his spark gap was influenced by light illumination. He noted this effect but did not investigate it in detail. Subsequent experimenters showed that when light is shined on some metals and metal oxides, electrons are emitted from the surface of the material. This effect explains Hertz's observation and is now called the *photoelectric effect.*

Although the details of the photoelectric effect are rather complex, the phenomenon can be easily understood qualitatively. We have already explained how electrons can be emitted from materials when sufficient heat energy is provided for the electrons to break the bonds that bind them to the material. In the photoelectric effect the energy required to break the electron bonds comes from the light. The energy of the light absorbed by the material is converted into kinetic energy of electrons which allows some of the electrons to leave the material. The *phototube* is a device that utilizes this effect to convert light intensity into an electric signal. The construction of a phototube is similar to that of a vacuum tube diode (Figure 9-2). It also consists of a cathode and an anode sealed in an envelope. In the phototube, however, the cathode is made of a photoemissive rather than a thermally emissive material. When the photocathode is not illuminated, the space between the anode and the cathode is nonconductive and current cannot flow through the circuit. But when light is shined on the cathode, electrons are emitted from it and a current flows through the circuit.

The basic elements of a facsimile transmitting and receiving system are shown in Figure 9-3. The picture to be transmitted is mounted on a revolving drum which simultaneously also moves axially. Light from a bulb is focused into a small spot on the picture. The amount of light reflected from the illuminated element is proportional to the reflectivity of that portion of the picture. The brighter the element the more light is reflected from it. The reflected light is collected by a lens and shined on the surface of the photocathode. Since the number of electrons emitted from the photosensitive cathode is proportional to the light intensity, the current flowing through the phototube is proportional to the brightness of the picture element being illuminated. As the drum rotates, the whole picture is sequentially sampled and therefore the current through the tube follows the brightness pattern of the picture. The signal generated in this way is called the *video signal*. This video signal is then transmitted by wire or radio waves to the receiving station. At the receiver a photosensitive film is placed on a drum whose motion is identical to that of the drum at the transmitting station. The received video signal is applied to a light source, causing it to glow with a brightness proportional to the applied signal. The light is focused on the rotating film, producing at each point an exposure that duplicates the original picture.

Very soon after its invention the facsimile was widely used in the transmission of news photographs. In 1924 the Marconi Wireless Telegraph Company transmitted pictures across the Atlantic.

The facsimile system can transmit single pictures, but it does not respond quickly enough to transmit the changing images that are necessary to produce the illusion of motion. At least 25 full pictures per second must be displayed in order for the eye to perceive the sequence as continuous motion.

PRINCIPLES OF TELEVISION

The basic ideas of modern television broadcasting were first published in 1908 by A. A. Campbell-Swinton in *Nature* Magazine. However, it took more than 20 years before the technology was sufficiently developed to build a working system. As we have mentioned before, the key to image transmission is the conversion of the visual image into an electric video signal. This is done by a technique known as *scanning*. We have already seen an example of scanning in the facsimile system. The visual image is examined or scanned line by line, and at each point an electric signal proportional to the local image brightness is generated. As an example, we can consider the scanning of the letter A shown in Figure 9-4a. Let us scan the picture from left to right in ten lines. The video signal generated from this scanning is shown in Figure 9-4b. This video signal can now reproduce the original picture (Figure

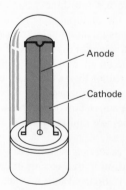

Anode

Cathode

Figure 9-2 Schematic
diagram of a phototube.

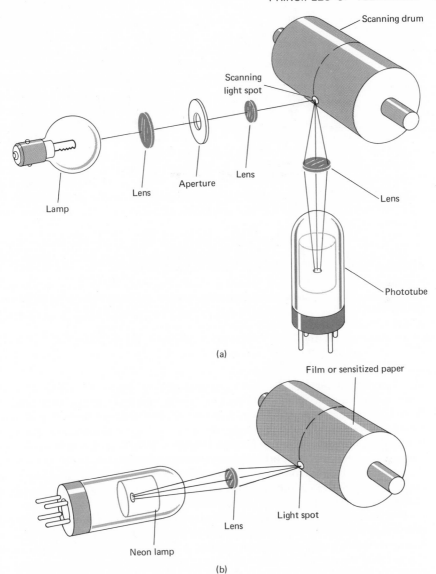

(a)

(b)

Figure 9-3 (a) Schematic arrangement for facsimile scanning. (b) Schematic arrangement for facsimile recording.

9-4c). Because of the finite size of the scanner which may be a light beam or, as we shall see, a beam of electrons, the reproduction is not as sharp as the original. Furthermore, because the scanning is done in a finite number of lines, the reproduced image is likewise made of the same number of lines. The smoothness and definition of the image can be improved by decreasing the size of the scanner and increasing the number of scanning lines.

The early attempts at scanning were all mechanical and were not very success-ful. In 1917 Vladimir Kosma Zworykin began to develop an electronic scanner

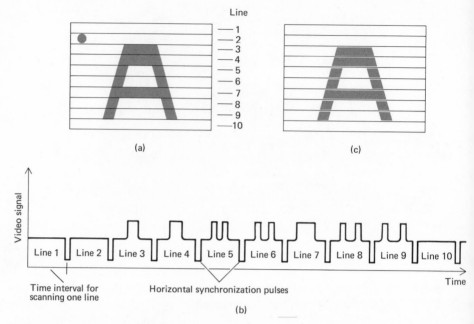

Figure 9-4   Scanning. (a) The letter A scanned by ten lines. (b) The resulting video signal. (The horizontal synchronization pulses are explained later.) (c) The reproduced image.

that was rapid enough for the reproduction of moving images. His initial work was done at the Russian Wireless Telephone and Telegraph Company. But after the Russian revolution Zworykin moved to the United States and completed his work at the Radio Corporation of America. The type of scanner developed by Zworykin is called *iconoscope*. The Zworykin television system was first demonstrated in 1929 at Rochester, New York. This system still had many problems to be solved before it could be used for commercial broadcasting. Television broadcasting began in the United States on July 1, 1941, but because of World War II, its initial development was slow.

The development of television in England was somewhat more rapid than in the United States. A British engineering group had developed at Electrical and Musical Industries in London an electronic scanner similar to the iconoscope. Commercial broadcasting was started by the British Broadcasting Company (BBC) in 1936, and when the service was closed in 1939 by the start of the war, 20,000 television receivers had been sold in London. The expansion of television broadcasting after World War II was immense. Today nearly all population centers are covered by television broadcasting.

Although the iconoscope scanner is now obsolete and has been replaced by better designs, it does illustrate clearly the basic method of fast scanning used in television transmission. A simplified diagram of an iconoscope tube is shown in Figure 9-5. It consists of an electron gun and a screen on which are deposited thousands of silver globules coated with cesium oxide. Both the gun assembly and the screen are enclosed in an evacuated envelope. The basic purpose of the electron gun is to produce a narrow beam of fast electrons for the scanning of the image on the screen. The operation of the gun is straightforward (Figure 9-6).

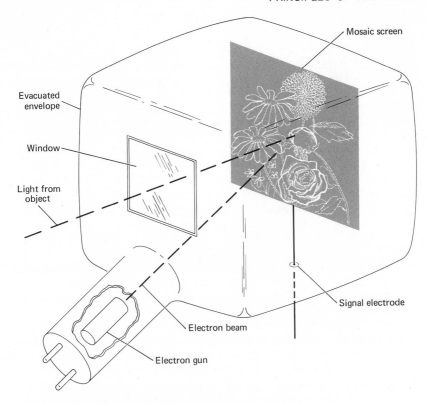

Figure 9-5 Simplified diagram of an iconoscope.

Electrons are emitted from a heated cathode and are then accelerated by a high voltage applied to a plate, *A*. This plate has a small hole in it which allows the electrons to pass through. The electron beam then travels through a set of mutually perpendicular plates that can deflect the beam from its original straight path. The beam can be moved up, down, or sideways by applying suitable voltages on the deflection plates. For example, if a positive voltage is applied to the *Y* plate, the

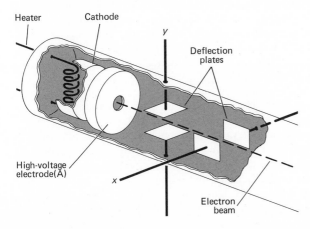

Figure 9-6 Schematic diagram of an electron gun.

electron beam will be attracted to the top plate and repelled from the bottom plate. The electrons are moving through this region so fast they do not actually hit the plate but continue toward the screen. The plates can then deflect the beam to any part of the screen. If the deflecting voltages are properly varied, the screen is fully scanned by the electron beam.

The globules on the iconoscope screens are sufficiently separated to remain electrically insulated from each other. If any one of these globules is illuminated by light, electrons are ejected from it due to the photoelectric effect and the globule becomes positively charged. The amount of charge on the globule is proportional to the light intensity. Thus when a visual image is projected on the screen, each globule becomes charged to a level proportional to the intensity of the image at that point. The narrow electron beam produced by the electron gun hits the screen and scans it horizontally line by line as shown in Figure 9-4. The horizontal deflecting plate causes the beam to move across the screen and trace out a line and the vertical plates cause the beam to jump to the next line. As the electron beam strikes the charged globule, a portion of the beam proportional to the charge enters the globule to neutralize it. This produces in the signal electrode a current that is proportional to the intensity of the image focused on the globules being scanned. This is the desired video signal. The electrons that are scattered from the screen are collected by an electrode; for the sake of simplicity this electrode has been omitted from the drawing in Figure 9-5.

In the commercial television systems used in North America the electron beam scans the whole screen in 525 lines. To convey motion the full scanning is repeated 30 times a second. A portion of the video signal which would result from scanning the letter A is shown in Figure 9-4. We note that a horizontal synchronization pulse is added at the end of each scan. This pulse controls the image reproduction at the receiving end. The video signal is used to amplitude modulate a carrier in a way that is identical to the method we discussed in connection with amplitude-modulation broadcasting. The carrier frequency is in the neighborhood of 50 megacycles, but each television transmitting station in a given locality uses a different carrier frequency. The carrier also contains the audio information associated with the television picture, which is usually transmitted by ordinary frequency-modulation techniques.

A block diagram of a television transmitter and receiver is shown in Figure 9-7. The roles of most of the components in the diagram are self explanatory. The modulated carrier containing the audio and video signals is launched into space by the transmitter antenna and is picked up by the receiving antenna. After demodulation the audio signal is applied to a speaker, and the video signal is converted into an image by a device called the *cathode-ray tube* or, simply, *picture tube*.

The origins of the cathode-ray tube can be traced back to the 1850s (Figure 9-8). It consists of an electron gun which is similar to that used in the iconoscope and a screen coated with a fluorescent material that glows at the point at which it is struck by an electron beam. The electron beam again scans the fluorescent screen line by line in synchronization with the scanning pattern at the transmitter. The motion of the beam in the cathode-ray tube is controlled by the synchronizing pulses that are sent along with the video signal. The amount of light emitted by the fluorescent screen is proportional to the number of electrons hitting it. The number of electrons in the beam is controlled by the video signal that is applied to the intensity control grid. A large video signal allows more electrons to pass through to the screen than does a small signal. In this way the image intensity of transmitted

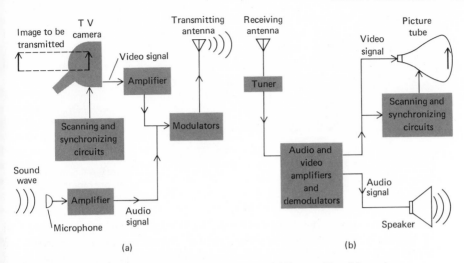

Figure 9-7   A simplified diagram of a television system. (a) Transmitter, (b) receiver.

picture is reproduced line by line on the receiving screen. Although the picture is built up line by line, the scanning is sufficiently rapid so that the eye responds only to the built-up image as a whole.

COLOR TELEVISION

Color television transmission began in the United States in 1954. It is based on the principle that any color can be reproduced by the proper combination of the

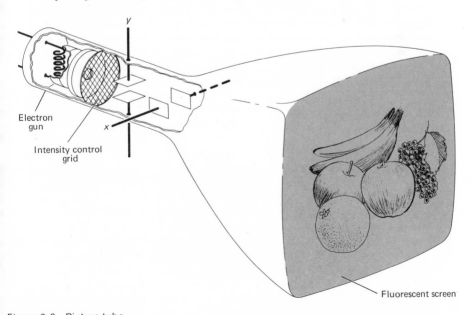

Figure 9-8   Picture tube.

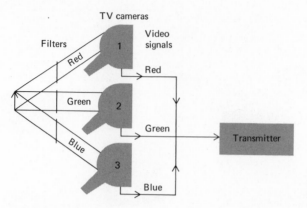

Figure 9-9 Simplified diagram of color television transmission.

three primary colors, red, green, and blue. At the transmitter the color content of the picture is separated into the three primary colors by means of mirrors and filters. Each of the three color images is projected on a separate television camera which produces a video signal proportional to the respective color content of the picture (Figure 9-9). At the receiver the television tube contains three electron guns, each of which is controlled by the video signal of one of the primary color patterns. The screen of the receiving tube is composed of three separate sets of phosphor patterns which produce, respectively, the three primary colors when bombarded by the electron beam (Figure 9-10). The phosphor patterns are uniformly distributed and the electron beam of each gun is so focused that it hits the phosphor corresponding to the color that controls it.

The focusing problems are more critical in color television than in black and white television transmission. The diameter of the phosphor dots is about 0.43 millimeter, so if the beam is that much out of focus it will hit the wrong colored phosphor and the picture will be distorted. In order to obtain sharper focusing the accelerating voltages associated with the color tube must be higher than in a black and white tube. If the tube is not designed properly, high-energy electrons may hit a tube surface and produce dangerous x rays. The emission of x rays is due to excited atoms which gained their energy from the bombarding electrons. These excited atoms release their energy in the form of electromagnetic radiation in

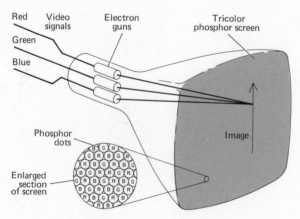

Figure 9-10 The tricolor television picture tube.

the x-ray wavelength region. In general, a color television is a much more delicate and sensitive instrument than the black and white set.

In our discussion of television we have omitted many of the details which are necessary to the proper operation of the system and which make the television a rather complicated instrument. It is truly a tribute to the efficiency of mass production that television sets can be manufactured at a price accessible to the general public. Oscilloscopes and other electronic instruments that are comparable in complexity to a television set but are not mass produced to the same extent cost 10 or 20 times as much as a television set.

10 Semiconductor Devices

The most important postwar technological events have been without a doubt the invention and development of semiconductor devices such as transistors, semiconductor diodes, and, more recently, integrated circuits. These devices perform the same functions as the conventional vacuum tubes, but they are so much smaller that they have made feasible completely new ventures. The modern computer contains so many components that its construction is possible only because of their small size. The advance in semiconductor electronics has been one of the cornerstones of the space program in which miniaturization is absolutely essential. Semiconducting devices are not only smaller but they also require much less operating power. This has given rise to new concepts in battery-operated appliances, such as radios, hearing aids, electric watches. Because semiconductor devices do not have heater filaments, they are ready for instantaneous operation. The possible applications of semiconductor devices are far from exhausted. We will certainly see their increased use in the fields of medicine, communication, and home appliances.

Actually semiconductor devices are not new. They were used at the very beginning of radio technology for the detection of radio signals. These were the early crystal detectors. Unfortunately these detectors were unreliable and their operation was not understood, so that when the vacuum tubes appeared, these semiconductor devices were abandoned. Work on the theory of semiconductors, however, continued and by the late 1930s the theory was reasonably well developed. But semiconductor materials were poor in quality and vacuum tubes were doing an adequate job, so there was very little impetus for developing semiconductor devices. During World War II a new problem arose, that of detecting radar signals. The conventional vacuum tubes could not respond to the high-frequency radar signals and again crystals had to be used to detect them. Renewed interest in these devices caused an increased effort to obtain pure semiconductor materials. By the end of the war pure samples of silicon and germanium were available and intense research on semiconductor devices was in progress. In 1948, shortly after the war, W.H. Brattan, J. Bardeen, and W.B. Shockley, working at the Bell Telephone Laboratories, invented and developed a practical transistor amplifier. For this work they received the Nobel Prize in 1956. The early transistors and other semiconductor devices were very expensive and delicate. Their performance and reliability were inferior to the vacuum tubes. But through the research and development that followed, semiconductor devices have been perfected to the point where they have replaced vacuum tubes in most applications.

THEORY OF SEMICONDUCTORS

Earlier we discussed very briefly the electrical properties of materials and we showed that some materials are conductors of electricity and others are insulators. There are also materials with electrical properties between these extremes called *semiconductors*. Our previous explanation of the electrical properties of matter was too superficial to explain the interesting properties of semiconductors. We will now examine in greater detail the interaction of atoms in matter and in this way obtain a more complete understanding of electrical properties. This subject is part of an important branch of physics called *solid state physics*. The interactions and phenomena we will describe are derived from quantum mechanics and have all been experimentally verified. Since the quantum theory necessary for the description of semiconductors is out of the scope of this discussion, our descriptions will deal only with the qualitative aspects of the subject.

Before we begin the discussion of matter, we must discuss the Bohr model of the atom in greater detail. You will recall that the atom consists of the positive nucleus and negatively charged electrons circling around it. Energy is associated with electrons as they move around the nucleus. The closer the orbit of an electron to the nucleus, the lower is its energy. In other words, if we want to increase the orbit of the electron, we have to transfer energy to the system. As we mentioned before an electron cannot occupy any arbitrary orbit about the nucleus; it can be found only in certain discrete orbits (sometimes called *allowed orbits*). For each orbit there is a maximum number of electrons that can occupy it. In most cases only the electrons in the outer orbit are of importance since the inner electrons remain close to the nucleus and are not involved in interactions with other atoms.

With each allowed orbital configuration of the electron, there is associated a specific amount of energy. Therefore, instead of speaking of the electron as being in a certain orbit, we can refer to it as having the corresponding amount of energy. Each value of energy that an electron can have in the atom is called an *energy level*. An energy level diagram for an atom is shown in Figure 10-1. Of course, every element has its own characteristic energy level structure. Note that the electron can occupy only specific energy states; that is, in a given atom the electron can have an energy E_1 or E_2, but cannot have an energy between these two values. This is a direct consequence of the restrictions on the allowed electron orbital configurations.

The lowest energy that an electron can occupy is called the *ground state*. This

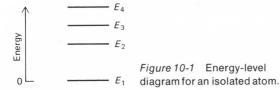

Figure 10-1   Energy-level diagram for an isolated atom.

state is associated with the orbital configuration closest to the nucleus. The higher allowed energy levels, called *excited states,* are associated with larger orbits and different orbital shapes. Normally the electron occupies the lowest energy level but it can be excited into a higher energy state by adding energy to the atom. This can be done with electromagnetic radiation at an appropriate frequency or by collisions with electrons and other particles. (See Chapter 12.)

This model of discrete atomic energy levels is applicable only when the atoms are isolated from each other. The situation is grossly altered when the atoms are so close together that their electrons interact. In a solid the atoms are held closely together. They interact so strongly that the individual atomic energy levels are no longer a useful concept. We must therefore examine what happens to the energy levels in such an interacting ensemble of atoms. To simplify the problem let us first look at the interaction of only two identical atoms. The new set of energy levels applicable to the diatomic system as a whole can be calculated using quantum mechanics. Qualitatively the situation is as follows. Each atom brings to the composite system its own set of energy levels. In this case, in which the atoms are identical, the total number of energy levels of the diatomic system is doubled. However because of the atomic interactions, each energy level is shifted somewhat from the original, which results in the energy level structure shown in Figure 10-2. For each energy level of the individual atom, the diatomic system has two closely spaced energy levels. In the coupled system the outer electrons are no longer associated with the individual atoms. They belong to the diatomic system as a whole. These electrons are indistinguishable and can occupy any of the energy levels of the composite system. We have already pointed out that it is this sharing of electrons by atoms that binds the atoms together to form molecules. These ideas can be extended to a system of the many interacting atoms constituting a solid. Here again each atom brings its energy levels to the composite and all the energy levels are somewhat shifted due to the mutual atomic interactions. The energy levels that were originally associated with a given configuration of the isolated atoms are now clustered very close together forming an energy band rather than discrete energy levels. This situation is shown in Figure 10-3. As with the diatomic system, here also the interacting electrons are no longer associated with an individual atom. The electrons belong to the solid as a whole.

With this picture we can now explain the electrical conduction properties of solids. Let us consider the first two energy bands of the substance. For reasons that will become evident the lowest band is called the *valence band* and the upper band is called the *conduction band.* In an insulating material such as diamond all the electrons are in the valence band which is completely filled. This statement

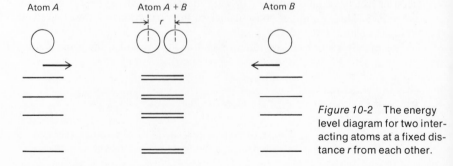

Figure 10-2   The energy level diagram for two interacting atoms at a fixed distance *r* from each other.

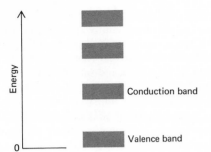

Figure 10-3   Energy bands in a solid.

can be explained in a more physical way by considering the type of interatomic binding in diamond. Diamond is made of carbon atoms, each of which has four outer interacting electrons. In the diamond crystal the carbon atoms are held together by the pairing of electrons from adjacent atoms (Figure 10-4). These bonds are strong and the electrons are not free to move. This type of binding between atoms is called a *covalent* bond and the electrons partaking in the binding are called *valence* electrons. It is shown by quantum mechanics that the binding between the carbon atoms is complete; that is, no other bonds can be formed between the carbon atoms because all electrons are in the valence band, and the valence band is filled. Such a substance does not have free electrons to conduct electricity and is therefore an insulator.

On the other hand, in a conducting material such as copper, not all electrons partake in the valence binding. One electron from each atom on the average is free to move through the metal and is therefore in the conduction band. Since these electrons in the conduction band are free to move under the influence of an electric field, they can conduct electricity.

The binding of atoms in a semiconductor material such as germanium is similar

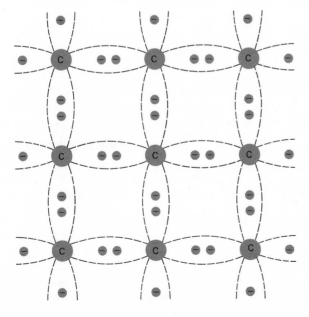

Figure 10-4   The interatomic bonding in diamond. The broken lines show bonding between atoms.

to the binding in insulators. Germanium has four outer electrons and each of these electrons shares the orbit with an electron from a neighboring atom. In fact at low temperatures germanium is an insulator. However, the covalent bonds in a semiconductor are not as strong as the bonds in an insulator. At room temperature the thermal energy is sufficient to break some of the bonds and set the electrons free. Since these liberated energetic electrons can move through the material, they are in the conduction band and can therefore conduct electricity. In germanium at room temperature there is approximately one such conduction electron for every billion germanium atoms. We recall that in copper there is conduction electron for every atom. However, the number of germanium atoms is so large that even this small fraction of conducting electrons produces a reasonable conductivity. At higher temperatures more of the valence bands are broken, thus increasing the number of electrons in the conduction band. The conductivity of semiconductors therefore increases with temperature.

For every electron that has been promoted into the conduction band there remains an incomplete bond in the valence band. This incomplete bond is called a *hole*. In the valence band a bound electron adjacent to the hole can move into it, leaving its original bond empty in return and thereby producing a new hole. When an electric field is applied to the semiconductor, current is conducted not only by the conduction electrons but also by the holes in the valence band. A very good analogy illustrating this concept was given in *Scientific American* in an article by Morgan Sparks. It is shown in Figure 10-5. Here the upper and lower floors of a garage are analogous to the conduction and valence bands in a solid. The cars represent the electrons. We can see that if the lower floor is fully occupied, corresponding to a filled valence band, there can be no movement of cars. If, however, a car is moved from the lower floor to the upper floor, motion of cars is then possible on both floors (Figure 10-6).

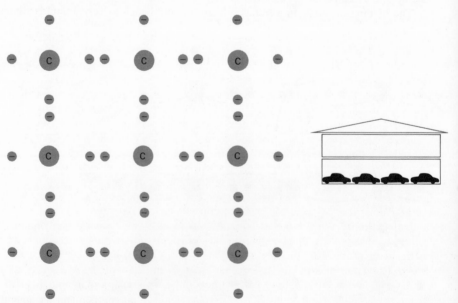

Figure 10-5 Insulators. In an insulator such as diamond the electrons are bound so tightly that they cannot conduct electricity. In the garage analogy this is shown by the cars completely filling the first floor permitting no traffic.

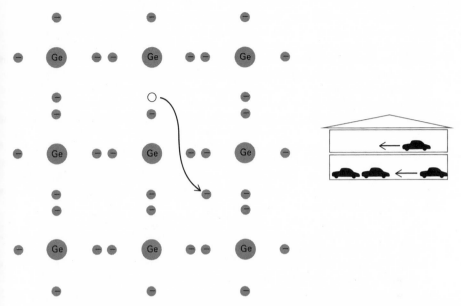

Figure 10-6 Semiconductors. In a germanium crystal the electrons are not bound as tightly as in diamond. An electron may break away from its bond leaving behind a hole. Current can be conducted both by the electron and the hole. In the garage analogy this corresponds to moving a car from the first to the second floor. Traffic is now possible on both floors.

n-TYPE AND *p*-TYPE SEMICONDUCTORS

In a pure semiconductor material such as germanium or silicon, the number of electrons in the conduction band is always equal to the number of holes in the valence band. This equilibrium can be altered in a controlled way by diffusing or melting into the semiconductor small amounts of arsenic or boron. The effects of these two elements on the electrical properties of the semiconductor are very different. The infusion of arsenic increases the number of electrons in the conduction band, whereas the addition of boron increases the number of holes in the valence band.

A semiconductor in which the number of conduction electrons has been increased is called *n-type*. A semiconductor with an increased number of holes is called *p-type*. It is the difference in the electrical properties of *p*- and *n*-type materials that is the basis of operation for most semiconductor devices.

The increase in the number of conduction electrons produced by the addition of arsenic is due to the electronic structure of the arsenic atom. Arsenic has five outer electrons. When a small amount of arsenic is melted into germanium (or silicon), the arsenic atoms become bound to the germanium atoms; the neighboring germanium atoms can provide only four valence electrons to bind the valence electrons of the arsenic atom. In other words, only four of the five electrons in the arsenic atom participate in valence bonding. This leaves one unbound electron from each arsenic atom to move freely in the material. This electron is in the conduction band (Figure 10-7).

The addition of boron has the opposite effect. It increases the number of holes. Boron has only three valence electrons. Thus when boron is added to the ger-

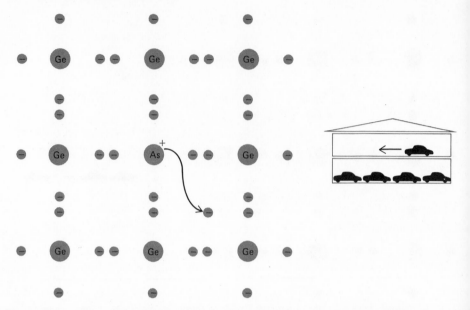

Figure 10-7   *n*-type semiconductor. Arsenic atoms have five outer electrons. When an arsenic atom is added to a germani n crystal four of its electrons are bound in the crystal structure and the fifth electron is free to move within the crystal and conduct current. In the garage analogy this corresponds to placing a car on the second floor which permits traffic there.

manium matrix, one out of the four possible bonds with the germanium atoms remains unfilled. This unfilled bond becomes a new hole in the valence band (Figure 10-8).

SEMICONDUCTOR DIODE

A semiconductor diode is formed by infusing arsenic into one end and boron into the other end of a semiconductor crystal. The resulting *p*- and *n*-type materials meet in a region called a *p-n junction*. A semiconductor diode with its schematic symbol is shown in Figure 10-9. The semiconductor diode is functionally identical to the vacuum tube diode; it passes current in one direction only. A current can flow through the semiconductor diode if the *p*-type material is positive with respect to the *n*-type material (Figure 10-10a). If the polarity is reversed, current cannot flow (Figure 10-10b). This unidirectional behavior is due to different electrical properties of the *p*- and *n*-type materials.

The *p*-type material is positive in the conducting mode; therefore electrons flow from the conduction band of the *n*-type material into the empty conduction band of the *p*-type material. At the same time the bound valence electrons from the *n*-type material also move into the holes in the *p* side. The electrons that leave the *n* side are replaced by others arriving through the closed circuit. A current is thereby conducted through the diode. On the other hand, if the *p*-type material is made negative, the electrons in the conduction band of the *n*-type material are forced to move away from the *p* side. But this initial surge cannot continue because there are

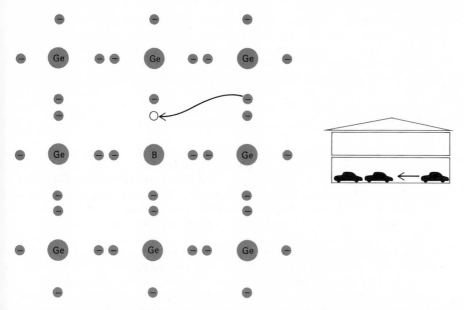

Figure 10-8 p-type semiconductor. Boron atoms have three outer electrons. When a boron atom is added to a germanium crystal, one empty bond results for each boron atom. Adjacent electrons can move into this hole and conduct electricity. In the garage analogy this corresponds to producing an empty space on the first floor which permits traffic on that floor.

no electrons in the conduction band of the *p*-type material to replace those that have moved away from the junction; current flow via conduction electrons is therefore blocked. Similarly, since there are no holes in the valence band of the *n*-type material, valence electrons cannot flow from the *p* to the *n* side. Thus when the *n* side is negative with respect to the *p* side, current flow is blocked through both channels and the diode behaves as an insulator.

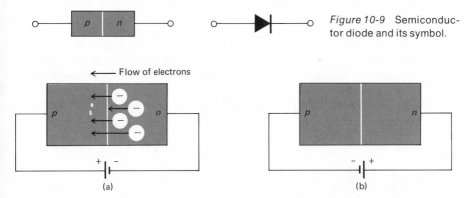

Figure 10-9 Semiconductor diode and its symbol.

Figure 10-10 (a) When the *p* side of the diode is positive with respect to the *n* side, both conduction and valence electrons flow from the *n* to the *p* side. (b) Current flow is blocked when the voltage connection is reversed.

The semiconductor diode can perform all the functions we mentioned in connection with vacuum tube diodes, but it has a number of advantages over the older device. Semiconductor diodes can be made microscopically small. They do not require any operating power or warm-up time. In addition, their frequency response can be made much higher than the response of vacuum diodes.

TRANSISTOR

The transistor is analogous to the triode vacuum tube. As shown in Figure 10-11, the transistor consists of a thin *p*-type material sandwiched between *n*-type materials. The three sections are called the *emitter,* the *base,* and the *collector* and are analogous to the cathode, grid, and plate, respectively, in a vacuum tube triode. As with the triode, transistor amplification is possible because a small signal applied to the base of the transistor can control a large current flow between the emitter and the collector.

The biasing arrangement for a transistor amplifier is shown in Figure 10-12. The power supply V_1 biases the collector positive with respect to the base. From our discussion of diodes we know that with this polarity current cannot flow from the base terminal to the collector. Another power supply, V_2, makes the base positive with respect to the emitter. This biasing causes a current to flow from the emitter into the base material. Now, what follows is the principle of transistor amplification. We might expect that the electrons flowing out of the emitter into the base will proceed out of the base material through the base terminal *A*. This is *not* the case. The base region is made so thin that most of the electrons get attracted into the positively biased collector before they can leave through the base terminal. But the base has a large effect on the size of the current flowing directly from the emitter to the collector. Small changes in the base voltage cause large changes in the collector current. This collector current flows through the output resistor, where it produces a voltage drop that is the amplified replica of the signal applied to the base. The operation of the transistor is therefore similar to that of the triode. In the triode the main flow of current is from the cathode to the plate, and this current is controlled by the voltage on the grid. In the transistor the main flow of current is from the emitter to the collector, and the current is controlled by the voltage applied to the base.

The transistor we have described is called the *n-p-n* transistor. Transistors can also be constructed with the emitter and collector made of *p*-type semiconductors and the base made of *n*-type material. Such a transistor is called the *p-n-p* transistor. Its operation is similar to the *n-p-n* structure except that the bias voltages are reversed.

Since the transistor requires biasing, it does consume some power, but less than a vacuum tube. About half a million transistors can be operated on the power required by one vacuum tube.

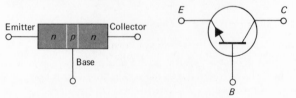

Figure 10-11   The transistor and its symbol.

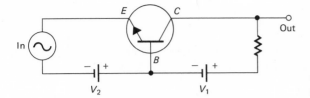

Figure 10-12   Transistor amplifier.

INTEGRATED CIRCUITS

During the last 10 years diffusion and photochemical techniques have been de-veloped that make it possible to produce on a single small semiconductor crystal complete circuits consisting of hundreds of individual components. These are called *integrated circuits*. Transistors, diodes, resistors, and capacitors can all be pro-duced on the crystal by forming microscopically small *p*-type and *n*-type regions. The tiny elements are then interconnected by metalizing appropriate portions of the crystal surface. A schematic illustrating the formation of two integrated tran-sistors is shown in Figure 10-13. Integrated resistors are produced by making con-tact through a single *p*-type region. Capacitors are obtained through oxide coating deposited on the surface. A photograph of an actual integrated circuit is shown in Figure 10-14. Such a circuit is usually enclosed in a small container with leads that can be connected to other components.

The small size of integrated circuits has profoundly changed electronic design concepts and feasibilities. This is especially true in computers in which the capacity is determined by the number of computer elements. The size diminution of an IBM computer component is shown in Figure 10-15. The early IBM computers using vacuum tubes could perform 39,000 additions in 1 second. The introduction of transistors in 1955 increased the computer capability to 204,000 additions per second. With the newest integrated circuits the computer can perform 15,000,000 additions in 1 second.

SOLAR CELL

A solar cell is a semiconductor device that converts light into electric energy. It is basically a semiconductor diode with a large junction area between the *p*- and *n*-type materials. When the *p-n* junction is illuminated by light, a current is produced by the cell (Figure 10-16). The details of the mechanism by which light is con-verted into current in the *p-n* junction are rather complex. The energy from the

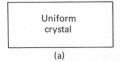

(a)

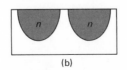

(b)

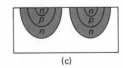

(c)

Figure 10-13 The formation of two transistors as part of an integrated circuit. Starting with a uniform crystal of germanium or silicon (a) successively *n*, *p*, and *n* materials are dif-fused into the crystal (b,c), forming in this case two transistors. The transistors can be interconnected by metalizing the crystal surface.

Figure 10-14   An integrated microcircuit used in radio and television. (Courtesy of the General Electric Company.)

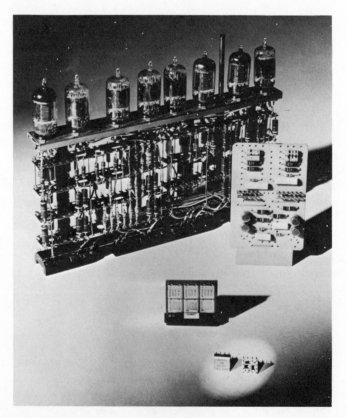

Figure 10-15   The size diminution of computer components. The circuit in the background was manufactured in the 1950s using vacuum tubes. The component in the foreground uses integrated circuits. (Courtesy of International Business Machines Corporation.)

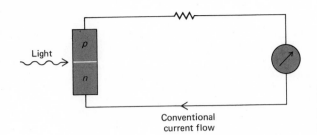

Light

Figure 10-16 Solar cell.
When light is shined on the
junction a current is generated.

Conventional
current flow

light breaks some of the valence bonds in the junction and excites the electrons in
the conduction band. This produces a flow of electrons through the external circuit
from the *n* to the *p* side of the cell.

The solar cell can convert about 10 percent of the light into electric energy.
Although this is a rather low conversion efficiency, the solar cell has poved economical for providing electric power from sunlight for satellites, remote telephone
amplifiers, and even portable radios. Solar cells are certain to be put into wider use
as they become cheaper and more efficient.

11 Communication at High Frequencies

The frequency of carrier waves used in communication has been getting progressively higher. At present many intercity and intercontinental communication links are made with carriers in the frequency range of kilomegacycles (10^9 cycles) per second. Radiation in this frequency range is called *microwave radiation* or simply *microwaves*. With the invention of lasers it has become possible to generate carriers with frequencies of 10^{15} cycles per second. Even though local radio and television broadcasting will probably continue with carriers in the megacycle frequency range, large-scale communication networks will certainly move to carriers with as high a frequency as is technically feasible.

BASIC CONSIDERATIONS

There are two important reasons for transmitting information with high-frequency carriers: (1) the information-carrying capacity of the carrier increases with frequency and (2) for point-to-point communication it is possible to aim high-frequency carriers and thus utilize a greater portion of the transmitted energy. As an example of the first reason, consider the problem of interurban telephone communication. It is certainly not economically feasible to provide a separate cable connection for each of the thousands of simultaneous conversations. A method must be devised for transmitting many individual signals at the same time along a single pathway. Such a technique has, in fact, been developed. It is called *multiplexing*. In multiplexing, a number of different signals modulate the same carrier, which is then transmitted from one city to the other. At the receiving terminal the carrier is demodulated and the separated signals are sent to their destinations. Of course, the carrier has to be modulated in such a way that the individual signals do not interfere with each other. To understand the problems involved in this technique we must reexamine the process of modulation in terms of the frequency changes it produces in the carrier.

You may recall that any wave pattern can be analyzed into its constituent sinusoidal components. For example, the wave shape shown in Figure 11-1a has the sinusoidal components shown in Figure 11-1b. The fundamental is at 10 cycles

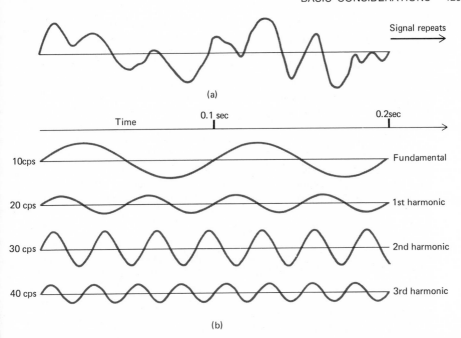

Figure 11-1   (a) A wave shape; (b) its frequency components.

per second. The first, second, and third harmonics are at 20, 30, and 40 cycles per second, respectively. The relative amplitudes of these components are shown in the figure. The sinusoidal components can be represented on a plot called the *frequency spectrum*. Such a spectrum for the wave shown in Figure 11-1 is plotted in Figure 11-2. This plot shows the frequency of the harmonic components and their relative amplitudes. The frequency spectrum is discrete, that is, it has separated frequency components. The components are at 10, 20, 30, and 40 cycles per second and there are no frequency components between these values. A discrete spectrum is characteristic of regularly repeating wave shapes. However, if the wave is not a regularly repeating function of time, it will contain sinusoidal components at all frequencies within a certain range. The lowest frequency is determined by the slowest changes in the wave, and the highest frequency component is determined by the fastest fluctuation present in the wave pattern. As an example, let us consider the frequency spectrum of a telephone signal. The frequency range of a tele-

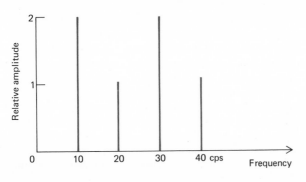

Figure 11-2   Frequency spectrum of wave shape in Figure 11-1a.

phone signal is from about 200 to 4000 cycles per second. A typical telephone audio signal pattern and its associated frequency spectrum are shown in Figure 11-3. Since the wave pattern is not regularly repeating, its frequency spectrum is continuous.

In the process of transmission the audio signal modulates a carrier by changing its amplitude or frequency. In order to understand multiplexing we must examine the effect of modulation on the frequency spectrum of the carrier. We will do this only for the case of amplitude modulation, but the ideas are also applicable to frequency modulation.

The frequency spectrum of the unmodulated carrier is simply a single point at the carrier frequency (Figure 11-4). This should be obvious since the carrier is just a single-frequency sinusoid. But the amplitude-modulated carrier is no longer a simple sinusoid. Since the amplitude of the carrier varies in accord with the audio signal, we must expect its frequency spectrum to be more complex. Actually the basic features of this spectrum can be derived from elementary trigonometry, but we will only present the results.

As shown in Figure 11-5 the frequency spectrum of the amplitude-modulated carrier consists of the original carrier frequency sandwiched between frequency components whose relative amplitudes are identical to the spectrum of the modulating audio signal. The frequency components above and below the carrier frequency are called upper and lower *sidebands,* respectively. Since each of the sidebands contains the full audio spectral information, only one of them needs to be transmitted (Figure 11-6). Thus from the point of view of the signal-frequency content, modulation is viewed simply as the shifting of the audio-frequency spectrum to the carrier-frequency region (Figure 11-7). Demodulation is just the reverse process in that the frequency components are shifted back down to the audio range (Figure 11-8).

Now we can explain the principle of multiplexing. Each of the audio signals to be transmitted by the main carrier first modulates its own local carrier (Figure 11-9). Since the frequency of each local carrier is different, the frequency spectrum of every audio signal is shifted by a different amount. The main carrier is then simultaneously modulated by all the local carriers. This results in a new frequency spectrum in which the individual frequency spectra are separated from each other and are placed side by side next to the carrier frequency. To achieve this the local carriers must differ from each other by at least the frequency width of each of the

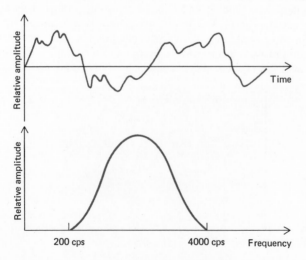

Figure 11-3   Waveshape of telephone audio signal and its frequency spectrum.

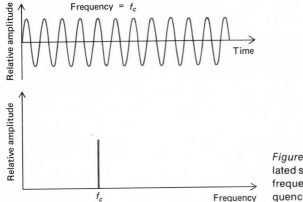

Figure 11-4 An unmodu-
lated sinusoidal carrier at
frequency f_c and its frequency spectrum.

audio signals. The modulated carrier is transmitted along a single pathway to its destination, where it is sequentially demodulated in a way that separates the audio signals.

In the process of multiplexing, the frequency of the carrier has been broadened in proportion to the number of audio signals that are multiplexed on it. If the carrier is, for example, at 1 megacycle and it is multiplexed by 100 signals, each with a 4000-cycle width spectrum, the frequency spectrum of the multiplexed carrier extends from 1 megacycle to 1.40 megacycles (Figure 11-10). Now we come to the main point. The number of individual signals that can be multiplexed on a carrier is limited by a practical but very important consideration. Each frequency region re-

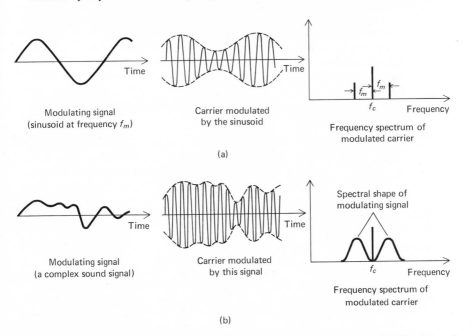

Figure 11-5 Amplitude modulation and the resulting spectrum for (a) modulation by a sine wave and (b) modulation by a complex signal.

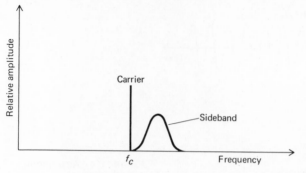

Figure 11-6   It is sufficient to transmit only one of the sidebands of a modulated signal.

quires different techniques for the generation, transmission, and detection of electromagnetic radiation. Since multiplexing broadens the frequency spectrum of the carrier in proportion to the number of signals being transmitted, this number must be limited so that the frequency components remain in the range for which the system was designed. From this consideration we will show that the higher the carrier frequency, the larger the number of individual signals that can be multiplexed on it.

To illustrate how the carrier capacity depends on the frequency range, let us assume that the highest frequency component produced by multiplexing is allowed to be no more than double the frequency of the unmodulated carrier; that is, if the unmodulated carrier is at 1 megacycle, the modulated carrier spectrum is allowed to reach 2 megacycles. Let us now compute the number of signals that can be

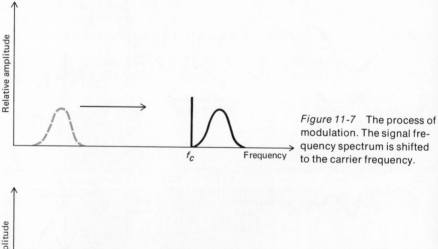

Figure 11-7   The process of modulation. The signal frequency spectrum is shifted to the carrier frequency.

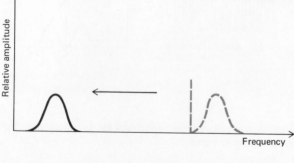

Figure 11-8   The process of demodulation. The signal-frequency spectrum is shifted back to its original position.

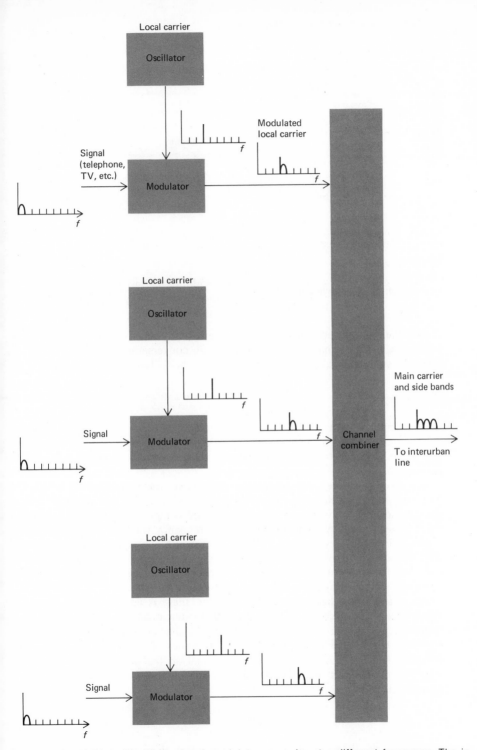

Figure 11-9 Multiplexing. Each signal modulates a carrier at a different frequency. The individual signals are then combined for simultaneous transmission on a single carrier.

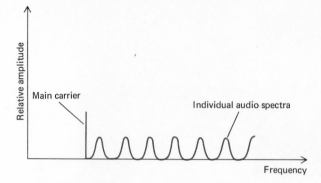

Figure 11-10 Frequency broadening due to multiplexing.

multiplexed in different frequency regions. This is obtained most simply from the equation

$$\text{number of allowed signals} = \frac{\text{allowed width of the multiplexed carrier}}{\text{frequency spectrum width of the individual audio signals}}$$

Thus if the multiplexed transmission is in the radio-frequency region between 1 and 2 megacycles, the number of 4000-cycle wide conversations that can be carried by this system is 250:

$$\frac{2 \times 10^6 - 1 \times 10^6}{4 \times 10^3} = 250$$

If the communication is in the frequency region between 1 and 2 kilomegacycles, the number of allowed conversations is 250,000. A further increase by a factor of 100,000 is obtained by going from microwave to optical frequencies (10^{14} cycles per second).

We have illustrated our discussion with voice communication in which the frequency-width requirement for each signal is a very modest few thousand cycles per second. The transmission of television and computer information requires a much larger frequency width than this. The frequency width of a television signal is about 6 megacycles. In fact, the frequency width of a signal increases directly with the complexity of the information it contains. With the long-distance information flow continuously increasing, the transmission capacity can be most economically increased by going to higher frequency carriers.

As stated at the beginning of this chapter, there is another reason for using high-frequency carriers in point-to-point communication. At high frequencies a larger fraction of the energy can be directly beamed from the transmitter to the receiver. This is due to the propagation properties of electromagnetic radiation. It can be shown by a rather complex wave analysis that when a beam of electromagnetic radiation is aimed into space, the beam remains nearly parallel for a distance given by

$$L = \frac{D^2}{4\lambda}$$

where D is the diameter of the source from which the wave is launched and λ is the wavelength of the radiation. Beyond this distance the beam spreads into a cone,

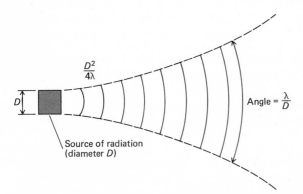

Figure 11-11 Spreading of a beam of radiation.

the angle of which is $\theta = \lambda/D$ (Figure 11-11). In this equation the angle θ is given in units of radians (1 radian is 57.3 degrees).

Let us first consider the propagation of radiation at radio frequencies, say at 1 megacycle. Here the wavelength is 300 meters and the size of a reasonably large antenna may be 10 meters. From the equations that describe the spreading of the wave we calculate that the radiation at this frequency spreads out immediately after launching, and therefore we conclude that it is not possible to launch a beam at this frequency (Figure 11-12a). The situation is somewhat better at microwave frequencies. At a frequency of 30×10^9 cycles per second the wavelength is 1 centimeter. With a 1-meter diameter antenna, the beam remains parallel for about 25 meters and diverges thereafter at an angle of 0.57 degrees (Figure 11-12b). A beam that started with a diameter of 1 meter spreads into a diameter of 100 meters at about 10 kilometers from the source. Although the diameter increases by only a factor of 100, the area of the beam increases by a factor of 10^4. The power per unit area decreases by the same factor. Therefore at a distance of 100 kilometers, a receiving antenna with a 1 square meter area can intercept only 1/10,000 of the initial transmitted power.

Point-to-point communication can be almost ideal with a radiation at the frequency of light which is obtained from laser sources. Here the wavelength is typically 10^{-4} centimeters and the beam diameter at the source is on the order of 1 centimeter. The laser beam remains parallel for about 0.25 kilometers and then spreads with an angle of 0.0057 degrees. At 10 kilometers the beam is spread into an area of about 1 square meter, but with suitable lenses placed about 1 kilometer apart, the spreading can be reduced to a point where nearly all the transmitted power is received by the detector (Figure 11-12c).

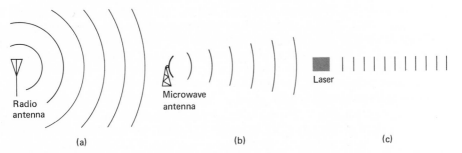

Figure 11-12 The spreading of electromagnetic radiation emitted by (a) a radio antenna, (b) a microwave antenna, and (c) a laser.

There are also some disadvantages to high-frequency communication. As the frequency is increased, the radiation is more and more susceptible to scattering and attenuation. While radio-frequency waves penetrate through fog, rain, and forests, high-frequency radiation bounces off these obstacles and is lost. This property of high-frequency radiation is in some cases very useful, but in communications it presents a problem.

Waves at radio frequencies can to some extent follow the curvature of the earth and can be reflected from the ionosphere. This allows direct communication over long distances. High-frequency radiation (higher than about 20 megacycles) does not bend around the earth nor is it reflected by the ionosphere. Direct over-the-horizon communication is therefore not possible with high-frequency carriers. This limitation is overcome by using relay stations or artificial satellite reflectors (see the following sections). In addition to the reception and transmission of the signal, in every communication system the signal has to be conducted to various circuit points within the transmitting and receiving equipment. It is difficult to conduct high-frequency signals along bent wires. For example, microwaves have to be conducted along specially constructed hollow metal pipes called *wave guides,* the size of which is on the order of the wavelength of the radiation. Wave-guide connections are expensive and much more cumbersome than the simple cable connections used in low-frequency circuits. However, in the overall picture of long-distance communication, the advantage is with high-frequency carriers.

MICROWAVES

It is interesting to note that Hertz's early experiments with electromagnetic radiation were done in the microwave-frequency region, about $0.10 - 100 \times 10^9$ cycles per second. This was not by choice. It simply happened that radiation produced by a spark of the type used by Hertz is in that frequency range. However, the pulses of radiation that were available to Hertz are unsuitable for modulated voice communication; for this the carrier must be continuous, have a stable amplitude, and be at a single well-defined frequency so that a receiver can be tuned to it.

With the development of vacuum tube oscillators it became possible to generate stable carriers for radio communication. But in this early stage of development only relatively low frequencies could be generated. Thus amplitude-modulation broadcasting developed in the 100-kilocycle frequency range. As the technology improved, carrier frequencies were extended into the megacycle range and this became the region of television and frequency-modulation broadcasting.

There is a practical limit to the maximum frequency that can be produced by a vacuum tube oscillator. For a vacuum tube to amplify a signal properly, the electrons in the tube must travel from the cathode to the anode in a time that is short compared to the duration of 1 cycle of the amplified signal. If the transit time is longer than the period of the signal, the electrons in the tube are influenced both by the positive and the negative part of the signal applied to the grid. Therefore the electrons are first accelerated and then decelerated, resulting in a cancellation of the grid control. Thus as the operating frequency of the tube is increased, its size must be decreased in order to keep the transit time sufficiently short. It is not practical to extend the operation of a conventional vacuum tube to microwave frequencies. In that range the tube would be so small that very little power could be extracted from it. Similar considerations of electron transit times in semiconductor materials show that the transistor is also unsuitable for high-frequency signal generation.

In the late 1930s two brothers, R. H. and S. F. Varian, developed a new device for amplifying and generating electromagnetic signals at microwave frequencies. This device, which is now called a *klystron* (from the Greek word meaning "waves breaking on beach"), is shown schematically in Figure 11-13. An electron beam is produced by the cathode and the accelerating grid. As shown in the figure the electron beam passes through two structures marked output and input resonant cavity. The resonant cavity is a metallic can with two holes to allow the passage of the electron beam, and another hole through which microwave radiation is introduced into or extracted out of the cavity. The size and shape of the cavity are such that inside it microwave radiation at a specific frequency can be preferentially built up to very high energy levels. In other words, at microwave frequencies the resonant cavity behaves like the resonant capacitor-inductor circuit at radio frequencies.

As the electron beam passes through the input cavity, it is acted on by the microwave field built up from the input microwave signal. This field oscillates sinusoidally in time, and therefore its direction with respect to the electron beam changes so that the beam is alternately accelerated and decelerated by the cavity field. In this way the uniform flow of electrons that enters the cavity is converted into bunches. The bunched electron beam is an alternating current at the frequency of the input signal. (The number of electrons passing a given point varies at the frequency of bunching.) As the bunched-beam current passes through the output resonator, it induces in the cavity a microwave field at the bunching frequency. Part of this energy is extracted from the cavity as the klystron output. Since the field in the output cavity induced by the bunched electrons is much stronger than the input signal, the microwave signal is amplified. The klystron amplifier can be converted into a microwave oscillator by feeding part of the output back into the input. This is the general method of converting any amplifier into an oscillator.

Although many other types of microwave generators are now available, the klystron is still widely used.

RADAR

The most rapid developments in microwave technology took place during World War II in connection with *radar,* a device that is used to detect objects by reflecting radio waves from them. (The word "radar" is an acronym for "*r*adio *d*etecting *a*nd *r*anging.") A schematic diagram of a radar system is shown in Figure 11-14. The radar dish antenna launches a beam of pulsed high-frequency radiation. If there is an object in the path of the beam, the radiation is reflected from it and is returned to the antenna where it is detected. Information about the size and the composition of

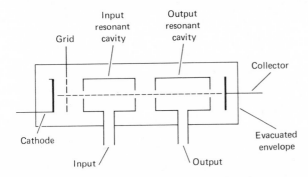

Figure 11-13   The klystron.

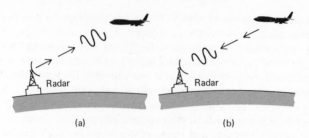

Figure 11-14 Operation of radar. (a) A pulse of radiation is emitted by the antenna. (b) The radiation reflected from the object is picked up by the antenna and detected by the radar receiver.

the object can be deduced from the magnitude of the reflected signal. The distance of the object from the transmitter is measured by the time interval between the transmission and the reception of a pulse. The individual pulses that are being sent out from the antenna are, of course, spaced far enough from each other to allow for the return of the echo from the most distant object of interest.

The antenna is rotated by a motor causing the radar beam to scan a wide region. The reflected signal is usually displayed on a specially designed cathode-ray tube, the surface of which is scanned by an electron beam that rotates in synchrony with the antenna. The intensity of the electron beam is controlled by the reflected radar signal so that the local brightness of the fluorescent screen is proportional to the intensity of the reflections from the corresponding point in space.

Radar was first patented in 1904 but could not be implemented until the high-frequency microwave sources became available. As we showed earlier only with high-frequency electromagnetic radiation is it possible to produce a narrow beam required for radar. Furthermore significant reflection occurs from objects only when they are larger than the wavelength of the impinging radiation. This is a general property of all waves that can be demonstrated by placing an object in the path of a wave propagating along a water surface. If the object is smaller than the wavelength (this is the distance between crests), the wave travels around the object without any reflections. The wave is reflected only if the object is larger than the wavelength (Figure 11-15).

Throughout the war great effort was made to extend radar operation to higher and higher frequencies. This was done mostly to keep ahead of the enemy's capability to produce signals that jammed and confused radar reception. But, in addition, the

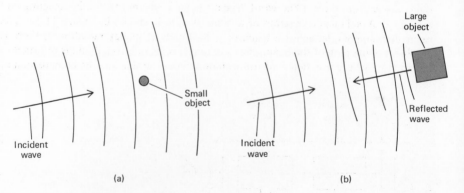

Figure 11-15 Reflection of a wave by an object. (a) Reflection of the wave is insignificant if the object in its path is smaller than the wavelength. (b) The wave is reflected if the object is larger than the wavelength.

higher-frequency operation made it possible to detect smaller objects and thereby improve the resolution of radar.

The improvements in microwave techniques led to the development of Doppler radar by means of which the speed of the detected object could be determined. Doppler radar is based on the principle that the frequency of reflected radiation is shifted from the original by an amount proportional to the motion of the reflecting object. This capability of radar has been indispensable to space technology. After the war, radar was put to civilian use in navigation, traffic control, and weather prediction.

MICROWAVE COMMUNICATION

Following World War II, microwave technology was applied to communication. Intricate microwave networks have been built in North America and Europe. The networks consist of microwave towers placed at 25- or 30-mile intervals. An antenna on top of the tower receives the signal, which is amplified and transmitted toward the next tower. The antenna is usually horn shaped or parabolic in order to focus the radiation onto the detector (Figure 11-16). At present most microwave transmission consists of telephone and teletype messages and television programs being

Figure 11-16 A microwave tower, part of Southern Railway's microwave system. (Courtesy of General Electric Company.)

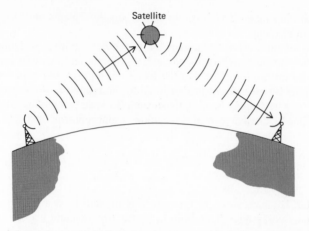

Figure 11-17   Microwave communication by satellite.

transmitted from a central source to the local stations. Usually the many individual signals are simultaneously transmitted on a single carrier, using multiplexing techniques. This type of tower microwave network is unsuitable for transatlantic transmission. Although it might be possible to build relay towers on floating, stabilized platforms, the cost for such a project would be unreasonably high.

The solution to an intercontinental microwave link was provided by space technology. The space program required techniques of communicating with satellites and space probes. These techniques were very quickly incorporated into the communication system. On August 12, 1960, the first communication satellite, Echo I, was launched. Echo I was simply a large balloon, 30 meters in diameter, which reflected microwave radiation back to earth (Figure 11-17). The radiation reflected back to earth spread into a large sphere and since it was not amplified, the signal detected by the receiving antenna on the earth was very weak. The Echo satellites did, however, demonstrate the feasibility of this type of communication.

The communication satellites following the Echo series have all been active (Figure 11-18). These satellites contain amplifiers to strengthen the signal and

Figure 11-18   A communications satellite, the Syncom Satellite, with its covering of solar cell paneling, is given a final checkout by technicians at the Hughes Aircraft plant in Culver City, California. (Courtesy of NASA.)

antennas to focus the signal preferentially back toward earth. The power for the amplifiers is obtained from banks of solar cells.

Dozens of satellites have already been launched into different orbits, some only a few hundred miles above the earth, others many thousands of miles into space. Almost from the beginning the communication satellites worked well and today transcontinental television, radio, and telephone transmission via satellite is routine.

The initial work on communication satellites was done in the United States. Soon, however, the Soviet Union launched its own communication satellites and now other countries are considering satellite systems for their own communication needs.

Satellite communication techniques are being used in areas other than the transmission of television and telephone signals. For instance, weather satellites transmit photographs of cloud formations, and military satellites are used for transmission of classified information.

As soon as microwave techniques became available, they were used in many laboratories for basic experiments with matter. Out of this work came the maser and eventually the laser which is the subject of the next chapter.

12 Lasers

The shortest wavelength of radiation that can be generated by a klystron is in the millimeter range. This limitation is imposed by the cavity size which has to have dimensions on the order of the wavelength for which it is designed. Since it is difficult and impractical to manufacture components smaller than a few millimeters, it appeared in the 1950s that the upper limit in the generation of high-frequency electromagnetic radiation had been reached. And then in 1960 the laser was developed. With the laser it became possible to generate radiation with frequencies 10,000 times higher than the highest microwave frequency. Radiation in this frequency range is light. Lasers were not developed specifically for communication; they grew out of the basic research on the interaction of radiation with matter. But once invented, lasers produced a great deal of interest in the communication field.

The generation of light is certainly not new. Efficient light sources have been available since the end of the last century. Light from a laser, however, has some unique properties that differentiate it from conventionally generated light and make it suitable as a carrier of information. In order to understand the nature of laser light, we must first examine the properties of light obtained from conventional sources.

The two types of light sources that have been used for many decades are the incandescent lamp and the spectral lamp. In an incandescent light source, such as the common light bulb, a filament wire is heated by a current to a very high temperature. As a result of heating, the electrons in the wire vibrate and emit electromagnetic radiation. Since the electrons in the wire are only loosely bound to the atoms, they can vibrate over a wide frequency range; that is, in a heated wire, we find electrons vibrating at all frequencies. Therefore the radiation emitted by the electrons is distributed over a wide frequency range. Furthermore, the radiation is emitted uniformly in all directions (isotropic radiation) (Figure 12-1). The heating of the wire also produces vibrations of the nuclei; but because the nuclei are much heavier than the electrons, their vibration is much smaller and the radiation emitted by them is negligible. The peak of the frequency distribution (spectrum) of the radiation emitted by the electrons is determined by the filament temperature. At lower temperatures, say 800°C, most of the radiation is at long wavelengths and only a small

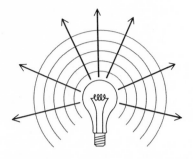

Figure 12-1 Light emitted by an incandescent light bulb propagates out in all directions and has a wide frequency (color) spectrum.

amount of the radiation is visible. The wire therefore appears to have only a faint glow. As the temperature of the wire is increased the motion of the electrons becomes more rapid, causing more of the emission to be in the visible range.

In spectral lamps the sources of radiation are the atoms in the lamp. The emitted light is at one or more discrete frequencies or colors. The details of this atomic light emission will be discussed shortly. Now we just note that the light is emitted by the atoms which have been excited to higher energy states. As the atoms return to lower energy states, the excess energy is emitted as light. The wavelength of the light is therefore characteristic of the energy states in the atom which are different for each element. Thus, for example, a neon lamp emits orange light and a sodium lamp emits yellow light.

The frequency of the light emitted by the atom is called its *resonant frequency*. The light emitted by a spectral source has a narrow frequency band (it is monochromatic) and is emitted in all directions. Although all atoms of a given element emit radiation at the same frequency, the emission from the atoms is random; that is, the atoms do not emit radiation at the same time and therefore the phases of the light emitted from different parts of the lamp are uncorrelated. Such a light source is called *incoherent* (Figure 12-2).

Actually the energy levels in an atom are not as sharply defined as we have indicated in our energy level diagrams. Because of collisions with neighboring atoms and other perturbations, there is a small variation in the energy of the levels. As a result the radiation emitted by atoms in the ensemble is not all at exactly the same frequency. The frequency of the emitted radiation is centered around the resonance frequency, but light is emitted both above and below this frequency. Thus the emitted light has a frequency width. Incidentally, no matter what the source, electromagnetic radiation always has a finite frequency width. It can be shown theoretically that neither atoms nor man-made oscillators can generate a signal at exactly a single frequency. The frequency deviations and instabilities can be very small, as in fact they are with atomic sources, but they are always present.

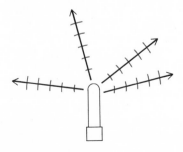

Figure 12-2 Light emitted by atoms in a spectral source has a narrow frequency range but it is emitted in all directions. The emissions from different parts of the lamp are uncorrelated.

Light from a laser is in some ways similar to that emitted by a spectral lamp. The sources of radiation are again atoms excited to higher energy states. The light is thus monochromatic and the frequency is characteristic of the atoms used in the laser. But whereas in a conventional spectral lamp the emission from the atoms is random, the emission in a laser is correlated. In other words, the atoms in a conventional lamp emit radiation as independent sources with randomly varying phases. On the other hand, the atoms in a laser emit light in a coordinated way so that the whole ensemble of atoms behaves as a single unit. Radiation from such a source is called *coherent* (Figure 12-3).

The most remarkable characteristic of the laser light source is the narrow beam into which the light is emitted. The light, which in a conventional spectral source is emitted in all directions, is now concentrated into a narrow parallel beam. Although it is possible to obtain a parallel beam of light from a conventional light source by using lenses or mirrors, only a very small fraction of the total light output can be channeled into the beam (Figure 12-4). In a laser the radiation from nearly all the atoms goes into the beam. Because the atomic emission is correlated, the phase of the radiation is the same across the whole beam and the frequency width is narrower than from an ordinary spectral source. These two properties of laser light, namely, the large power in a parallel beam and the highly monochromatic nature of the radiation, make the laser suitable for carrier communication.

BASIC PRINCIPLES OF LASERS

In order to explain the operation of the laser we must return to the atom and examine its interaction with electromagnetic radiation. Since the interaction of radiation in the frequency region of light is only with the outer electrons, we need not consider the inner-shell electrons. (Radiation at high frequencies, in the X-ray region, does interact with the inner electrons, but this is not in the scope of our discussion.) The energy level diagram for an atom is again shown in Figure 12-5. As we mentioned in our previous discussion an atom can be in any one of its energy levels, but in its unperturbed state it occupies the lowest level, which is called the *ground state*. In order for an atom to be promoted to a higher energy level, the required energy must be added to it from some external source. The addition of energy to the atom usually changes the electron orbital configuration to one with the higher energy.

An atom can be excited from a lower to a higher energy state in a number of different ways. The two most common methods of excitation are electron impact and absorption of electromagnetic radiation. Excitation by electron impact occurs most frequently in a gas discharge. If a current is passed through a gas of atoms, the colliding free electrons transfer some of their energy to the atoms. In this process the colliding electron is slowed down and the electron in the atom is promoted to a higher energy configuration. When the excited atoms decay back into the lower energy states, the excess energy is given off as radiation which produces the characteristic discharge color (Figure 12-6).

Interaction of atoms with electromagnetic radiation is more complex. The atom

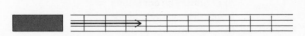

Figure 12-3 Light from a laser is emitted into a collimated beam and it is monochromatic.

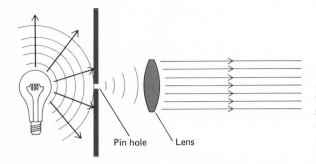

Figure 12-4 Parallel light from a conventional light source. Only a small fraction of the total light can be channeled into a parallel beam.

Pin hole   Lens

behaves as a resonant system. Only radiation at a specific frequency can cause transition from a lower energy level E_1 to a higher level E_2. This frequency is given by

$$f = \frac{E_2 - E_1}{h} \qquad (12\text{-}1)$$

The symbol h in this equation is called the *Planck constant* after the German scientist Max Planck (1858-1947), who first postulated this relationship between energy and frequency. If the energies E_1 and E_2 are measured in joules, then the value of h is 6.625×10^{-34}.

Every atom has more than one resonance frequency. There is a resonance frequency for every pair of energy levels. Thus an atom in state E_1 will absorb electromagnetic radiation at frequencies $(E_2 - E_1)/h$, $(E_3 - E_1)/h$, and so on, and in the process it will be excited into the corresponding energy state. But electromagnetic radiation at nonresonance frequencies will pass through the atoms without producing any change in them.

The amount of energy absorbed from the resonance beam by one atom being excited from state E_1 to E_2 is, of course, $E_2 - E_1$. From Equation (12-1) we can see that this amount of energy is equal to hf, where f is the resonance frequency. The quantity hf is called a *quantum* of energy.

One of the cornerstones of quantum mechanics was the discovery by Max Planck in the early part of this century that the energy content of an electromagnetic wave is not continuous. The energy of the radiation is in bundles each of size hf. This is exactly the energy of a quantum required to induce the resonant energy change in the atom. It is not possible to divide the energy of electromagnetic radiation at frequency f into units smaller than a quantum. Whereas in classical physics electromagnetic radiation was viewed entirely as a wave phenomenon, in modern physics it has been necessary to consider radiation as having at times particlelike proper-

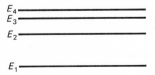

Figure 12-5 Energy level diagram for an atom. Under unperturbed conditions, atoms are in the ground state.

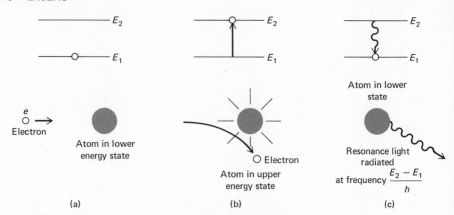

Figure 12-6   (a) When an electron collides with an atom, the atom may be promoted into an upper state (b). When the atom returns to a lower state, the excess energy is emitted as resonance radiation (c).

ties, with the particles being the individual quanta of energy; that is, in some interactions electromagnetic radiation behaves as a wave and in others as an assembly of particles. This view of radiation is known as the *wave-particle duality*.

After this dual approach to radiation was shown to be correct, it was extended to classical particles. It was shown that under some circumstances it is necessary to regard particles as having wavelike properties. One example which we have already mentioned is the electron in an atom. We pointed out that it is not correct to consider the electron as having a discrete, well-defined orbit in the atom. It is more appropriate to think of the electron as a wavelike cloud spread around the nucleus. An interesting qualitative discussion of these quantum mechanical ideas is found in Gamow's book listed in the References at the end of this book.

We are now ready to discuss the operation of lasers. Let us consider a group of atoms of a given element and let us focus our attention on two of the many energy levels. We will designate these levels A and B with energies E_a and E_b, respectively (Figure 12-7). The resonance frequency between these two levels is $(E_b - E_a)/h$. If there are some atoms in the lower state A, then radiation at the resonance frequency will be absorbed by the atoms and the atoms will be excited into state B. Now, what happens to atoms in energy level B? Eventually all the atoms will return to the lower energy state A since in an unperturbed system all atoms are in their ground states. There are two decay processes that are of importance in our discussion.

First, the atoms in level B tend to fall spontaneously to the lower energy level in a process that is analogous to objects falling off vibrating shelves. The excess energy is given off as a photon of resonance radiation. The radiation resulting from

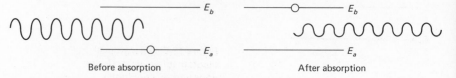

Figure 12-7   Atoms in lower state E_a absorb electromagnetic radiation at the resonance frequency $(E_b - E_a)/h$ and are excited into the upper state.

this process is called *spontaneous emission;* it is the source of light in resonance lamps. Spontaneous emission has the random properties we discussed in connection with resonance light sources (Figure 12-8).

Another way for the excited atoms to come down from the upper state is through interaction with resonance radiation. This process, called *stimulated emission,* proceeds as follows. Electromagnetic radiation at the resonance frequency interacts with the atoms in the excited state B. The interaction is such that the resonance radiation causes an atom in level B to make a transition to level A. The radiation emitted by the atom due to this process is in phase with the incoming resonance radiation. This means that the emitted radiation adds to the radiation that caused the emission. In other words, through the process of stimulated emission, the impinging radiation has gained energy from the excited atoms and thus has been amplified (Figure 12-9).

Now let us look at a realistic atomic system in which both states A and B are populated. This population may have been established by an electric discharge or by some other means. If light at the resonance frequency is passed through this ensemble, both absorption and stimulated emission occur. A very important point is that the probability for absorption and the probability for stimulated emission are the same; that is, the probability that an atom in the lower state absorbs a photon from the light beam is the same as the probability that an atom in the upper energy state is stimulated to add a photon to the beam. Therefore what happens to the beam as it passes through the atoms depends on the relative populations in the two states. If the population of the lower state is higher than the population of the upper state, more photons are absorbed from the beam than are added to it, and the net result is an attenuation of the beam (Figure 12-10a). On the other hand, if there are more atoms in the upper state, then more photons are added to the beam through stimulated emission than are subtracted from it by absorption, and the resonance light is in this way amplified as it passes through the atoms (Figure 12-10b).

In most cases the population of the lower energy state is larger than the population of the upper state since such a distribution is closer to the stable equilibrium distribution in which all atoms are in the ground state. For this reason a beam of resonance radiation is usually attenuated as it passes through an assembly of atoms. However, with some atoms and with suitable techniques it is possible to create and

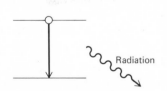

Figure 12-8   Atoms in the excited state can decay spontaneously into a lower state. The emitted radiation is called *spontaneous emission.* It is randomly emitted in all directions.

Figure 12-9   Electromagnetic radiation at the resonance frequency can stimulate the atom to make a transition to a lower state. The emitted light is called *stimulated emission.* It is in phase with the incoming radiation and adds to it.

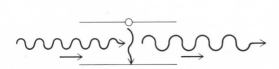

(a)   (b)

Figure 12-10 (a) If the population of the lower state is higher than the population of the upper state, resonance radiation is attenuated as it passes through the atoms. (b) If the population of the upper state is higher than the population of the lower state, resonance radiation is amplified as it passes through the atoms.

maintain a situation in which there are more atoms in an upper energy state than in the lower state. Such an atomic system amplifies resonance radiation and can be converted into an oscillator at this frequency. Such a device is called a *laser*, which is an acronym for "*l*ight *a*mplification through *s*timulated *e*mission of *r*adiation" (Figure 12-11).

As was the case with conventional amplifiers, the light amplifier is converted into an oscillator by introducing feedback. In lasers the feedback is done with mirrors. The amplifying laser medium is placed between two mirrors that are specially designed for the resonance wavelength. The oscillation process is started by random radiation originating from thermal or spontaneous emission. This radiation passes through the medium and is amplified. At the end of its passage the light is reflected by the mirror and again traverses the amplifying medium. Through repeated reflection the light continues to grow in intensity until it is limited by some saturation process. This usually occurs when the excited atoms are stimulated to emit light at the rate at which they are being produced.

The stimulated emission in a laser proceeds so rapidly that very few of the atoms have a chance to decay spontaneously before they are stimulated to emit their energy. Thus the energy of the excited atoms which in the absence of laser

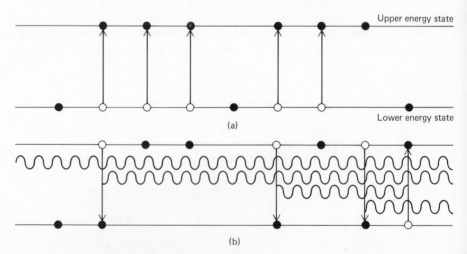

Upper energy state

Lower energy state

(a)

(b)

Figure 12-11 The laser medium. (a) An external excitation source produces a larger atomic population in the upper energy state than in the lower state. (b) When resonance radiation passes through this medium, both stimulated emission and absorption occur. However, since there are more atoms in the upper state than in the lower state, stimulated emission dominates and therefore the incident radiation is amplified.

amplification would go into isotropic spontaneous emission is channeled into the amplified beam propagating between the mirrors. Since all the excited atoms are being stimulated by the same beam, the emission is coordinated and the emitted light is coherent. In an actual laser the end mirrors are made partially transparent so that part of the light generated in the laser medium passes out through the mirror. Because the amplification is axial along the laser tube, the output beam is parallel and narrow (Figure 12-12).

The wavelength of the laser beam is determined by the energy levels in the amplifying atomic system. In argon it is possible to produce laser amplification between levels with an energy difference corresponding to green light. Thus the argon laser beam is green. The zinc laser produces two wavelengths, yellow and red, and helium-neon laser produces red and infrared light. So far more than 200 wavelengths have been obtained with lasers and this number is still growing rapidly.

Our discussion of lasers has so far been illustrated with only atoms in a gas. However, the ideas are equally applicable to molecules, liquids, and solids, all of which have energy levels with which laser amplification can be produced. Solids and liquids are much denser than gases and therefore more atoms participate in the stimulated emission process. For this reason solid and liquid lasers are capable of much larger power outputs than gas lasers. On the other hand, because of strong interactions between closely spaced atoms, the energy levels in solids and liquids are much broader than in gases, and therefore the light output of solid and liquid lasers is not as monochromatic as the light from gas lasers.

PRACTICAL LASER SYSTEMS

There are a number of ways of producing a system in which an upper state has a higher population than a lower state. In the helium-neon laser the two gases are mixed in a discharge tube. The electrons passing through the gas produce a large number of very long-lived excited helium atoms in the state that we have called A in Figure 12-13. The excited helium atoms collide with the neon and transfer their energy to the neon atoms. Through these collisions, state B, which is closest in energy to the helium excited atom, is preferentially excited in neon. Since state C is not excited in this collision process, state B becomes more populated than state C. The laser oscillates at the resonance frequency of these two levels. A schematic diagram and a photograph of a helium-neon laser are shown in Figures 12-14 and 12-15, respectively.

In the solid state ruby laser (which incidentally was the first operating laser) the excess population is produced by illuminating a ruby rod with a strong flash of light. This light causes transitions from state A to state B (Figure 12-16). The atoms in state B decay spontaneously to state C so rapidly that the population of state C becomes larger than the population of state A. Laser action proceeds between these two levels.

The theory and technology necessary for developing a laser were available by

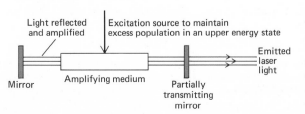

Light reflected and amplified

Excitation source to maintain excess population in an upper energy state

Emitted laser light

Mirror   Amplifying medium   Partially transmitting mirror

Figure 12-12   The laser. Two mirrors placed at the ends of the amplifying medium provide the feedback necessary for oscillation. The laser output is obtained through the partially transmitting mirror.

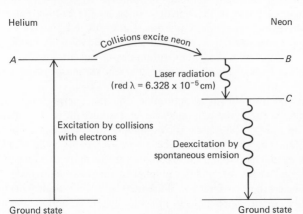

Helium

Neon

Collisions excite neon

A ──────────────────────── B

Laser radiation
(red λ = 6.328 x 10⁻⁵cm)

─── C

Excitation by collisions
with electrons

Deexcitation by
spontaneous emision

Ground state   Ground state

Figure 12-13 Potential energy-level diagram for the helium-neon laser. Electrons collide with helium atoms, producing excited helium in state *A*. Helium in this state collides with neon atoms. As a result of these collisions, neon is excited to state *B* and helium is deexcited to its ground state. An excess population is produced in state *B* with respect to state *C* of neon. Laser radiation is produced by stimulated emission from state *B* to state *C*.

the 1920s. As early as 1923 Tolman discussed stimulated emission, and in 1940 Fabrikant in the Soviet Union described in some detail the possibility of light amplification due to stimulated emission. These early ideas were not followed up and the laser was not built until 1960. The first successful application of amplification resulting from stimulated emission was at microwave frequency. In 1954 at Columbia University, Gordon, Zeiger, and Townes built a maser, a device that uses the same basic principle as the laser but operates at a microwave frequency. (Maser stands for "*m*icrowave *a*mplification through *s*timulated *e*mission of *r*adiation.") The original maser used ammonia molecules and it oscillated at 24×10^9 cycles per second. The maser grew out of the basic research done on the interaction of matter with microwaves and was not related directly to the earlier work. Since 1954 a number of other masers have been built; among them are the hydrogen, ruby, and rubidium masers. These have been most often used as frequency standards and as stable signal sources for experiments.

Proposals for a practical laser came in 1958 from Schawlow and Townes in the United States and Basov and Prokhorov in the Soviet Union. Very soon after this, laser experiments were being conducted in many laboratories, and in 1960 laser action was first observed by Maiman. This was with the optically excited solid ruby laser. In 1961 Javan, Bennett, and Herriott obtained maser oscillation in the helium-neon gas system, which is still one of the most useful laser systems. Townes,

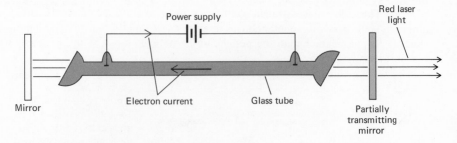

Power supply

Red laser
light

Mirror

Electron current

Glass tube

Partially
transmitting
mirror

Figure 12-14 Schematic diagram of a helium neon laser. A mixture of helium and neon gas is contained in a glass tube. The electrons for the excitation of helium are produced by the current passing through the gas.

Figure 12-15 A helium neon laser. (Courtesy of Bell Telephone Laboratories.)

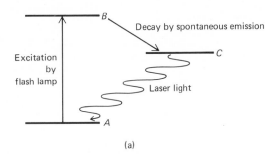

(a)

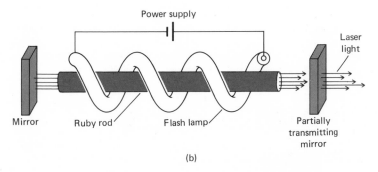

(b)

Figure 12-16 Ruby laser. (a) Light from the flash lamp excites atoms in the ruby rod from state *A* to state *B*. Atoms decay rapidly from state *B* to state *C*, producing an excess population in state *C* with respect to state *A*. (b) Schematic diagram of a ruby laser.

Basov, and Prokhorov were awarded the 1964 Nobel Prize in physics for their work in this field.

Lasers are now commercially manufactured by a number of companies. Depending on size and design they range in price from $200 to $20,000. Although most of the laser applications are still in the development stage, lasers have already been used successfully in welding, construction, and medicine. Because the laser beam is coherent and highly collimated, its total energy can be focused into a spot that is not much larger than the wavelength of light. With this focused energy very delicate drilling and welding can be performed. Focused lasers have been used in eye operations to fuse detached retinas and in skin operations to selectively burn off cancerous tissue.

In no field is the application of lasers pursued more diligently than in communications. Audio and television signals have already been transmitted with laser beams, but the systems are still in the experimental stage. Better techniques have to be developed for the modulation, detection, and transmission of laser beams.

With various electrical and mechanical methods it has recently become possible to produce very rapid deflection of a laser beam. The beam can be moved in a scanning pattern to display a large picture on a screen. The process is similar to the one used with electron beams. The scanning pattern is predetermined and the laser-beam intensity is controlled by the video signal. Because of the image-retention properties of the eye, the individually displayed lines on the screen merge into a total picture. Color picture reproductions have been produced by scanning the screen simultaneously with three primary color lasers (Figure 12-17). The pictures produced by laser scanning are larger and are much better resolved than those obtained with electron beams. At present, however, laser display systems are far too expensive for home use.

THE HOLOGRAM

Finally, in connection with lasers we will briefly describe holography, a new photographic method by which the full three-dimensional features of an object can be stored on film and then reproduced. Since the details of holography are rather complicated, we can give here only a general qualitative description of the technique.

Let us consider the conventional image formation from an illuminated object. Reflected light from every point on the object spreads out into the surrounding space (Figure 12-18). If a film is placed in the path of this expanding wavefront,

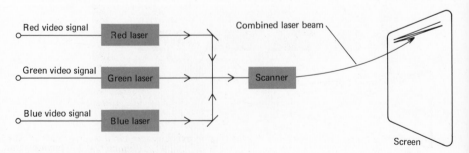

Figure 12-17   Laser color television.

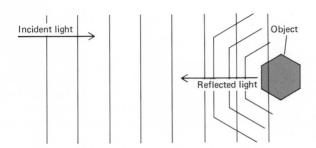

Figure 12-18 Reflected light spreading from an illuminated object.

it will become exposed. There will be light and dark regions on the film but the pattern will look nothing like the original object. In order to produce a recognizable image of the object, the spreading light has to be intercepted by a lens (or a pinhole) and focused onto the film (Figure 12-19). In this imaging process there is a point-to-point correspondence between the object and its image. When light is shined through a transparency made in this way, the eye sees a two-dimensional image of the original object (Figure 12-20).

Now let us return to the case in which the film without any lens is placed in the path of the expanding wavefront. The wavefront is made of the superposition of all the reflections from the individual points on the object. There are two parameters which fully describe this wavefront: the intensity of the light at every point along the wavefront and the relative phases among the portions of the light reflected from the different points on the object (Figure 12-21). The film alone records only the amplitude information (Figure 12-22a). If a beam of light is passed through this transparency, a meaningless jumble of light and dark patterns is formed. In order to capture the full wavefront, the phase information must be incorporated into the film. This is done by exposing the film simultaneously to the light reflected from the object and to a separate reference beam of coherent light (that is, light with a well-defined phase). The reflected light and the reference beam interact on the film to form the interference fringes shown in Figure 12-22b. These fringes record the phase of the reflected wavefront. On the portions of the film where the waves of the reference beam and the reflected light arrive in phase, the amplitudes of the two waves add to produce a light intensity that is greater than that of the individual waves. This produces the light fringes on the interference pattern. In the regions where the waves arrive out of phase, they cancel each other, causing a net reduction in the light intensity which results in the dark fringes in the interference pattern. A film that

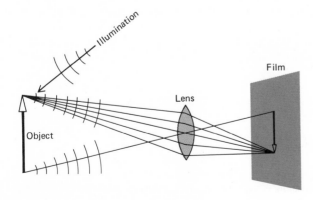

Figure 12-19 Conventional image formation.

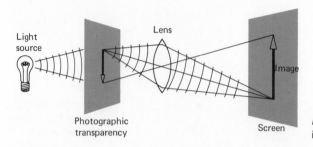

Figure 12-20 Conventional image reproduction.

records in this way both the amplitude and the phase of the wavefront is called a *hologram.*

In order to produce a hologram, both the illuminating light and the reference beam must be coherent and monochromatic. Only if this is the case can an interference pattern be formed on the film. Clearly, if either of the light waves has a randomly varying phase, the ordered reenforcement and cancellation of the waves will not occur. In addition, the light must have a relatively narrow frequency band; otherwise the interference patterns at various frequencies jumble the hologram. Still it is possible to record with holograms the color composition of the object. This is done essentially by forming separate holograms with the three primary colors.

The apparatus for making and reproducing holograms is shown in Figure 12-23. A coherent beam of radiation, usually from a laser, is split into two parts. One part illuminates the object and the other part is used as the reference beam. The hologram is formed on the film by the interaction of the reflected light and the reference beam.

When a coherent beam of light is passed through the hologram, the wavefront that was originally intercepted by the film is set into motion again. It is in every way identical to the original wavefront. If the wave is intercepted by the eye, the image of the object appears on the other side of the hologram as if it were viewed through a window. This image is a completely faithful three-dimensional replica of the original object. If the viewing direction is changed, a different perspective of the picture is revealed. The size of the film determines the extent of intercepted wavefront. Therefore the larger the hologram, the greater is the angle over which the image can be viewed.

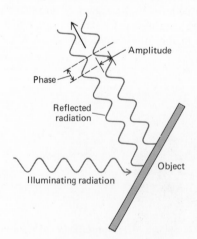

Figure 12-21 Both phase and amplitude must be recorded in order to reproduce a wavefront.

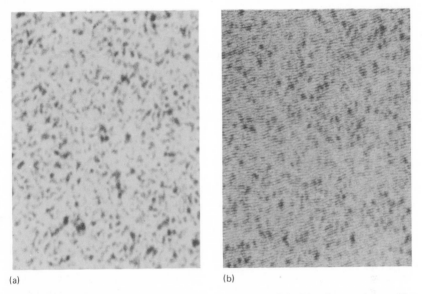

(a)   (b)

Figure 12-22   Photos of film (a) without reference beam and (b) with reference beam. (Courtesy of Perkin-Elmer.)

In the production of a hologram every point on the film receives light from all parts of the object. Therefore an exposed hologram can be cut into smaller sections and yet each section will reproduce the image. Naturally the angle through which the image can be viewed is decreased in proportion to the size of the hologram.

Holography was invented in 1947 by Dennis Gabor at the Imperial College of Science and Technology in London. The early work in this field was made difficult

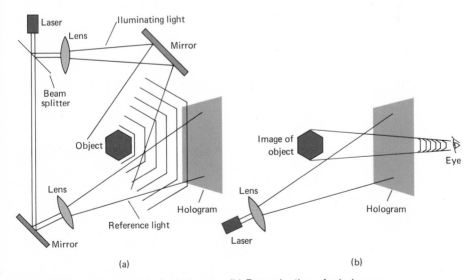

Figure 12-23   (a) Production of a hologram. (b) Reproduction of a hologram.

by the lack of coherent light sources. Once laser light sources became available, the techniques of holography were greatly advanced. Although at present holography is used only in experiments, work is being done on holographic motion picture and television reproduction. Holographic television is not yet technically practical, but it is feasible in principle. A television camera could record the hologram, which could then be transmitted and reconstructed at the receiver. A coherent light source at the receiver would then display the image. Such a system may well be the ultimate in image reproduction. In 1971 Gabor received the Nobel Prize in physics for his work on holography.

13 Communications in the Future

Predictions about technology usually tend to be conservative. When the wireless telegraph was first put into operation, very few people could foresee the development of our television system out of that primitive technology. In 1957 when the first Sputnik orbited the earth, the landing on the moon still seemed in the very distant future. I expect that my forecasts will also be conservative. In fact, without going into the realm of science fiction, I cannot imagine any profound innovations in communications. After all, basically we now do have nearly complete communications. We can transmit with high fidelity both sound and visual images over long distances. What other information is there to transmit? It may be that some future systems will be able to transmit the scented and tactile properties of objects, but I cannot foresee how this can be done nor am I certain that this is a desirable objective. Our discussion therefore is going to concern itself mostly with an examination of the present trends in communication technology.

TRENDS IN COMMUNICATIONS

The most notable trend in communication systems of the last decade has been the miniaturization of electronic components. The development of various solid state devices has made it possible to construct relatively inexpensive small radios, televisions, computers, electronic eavesdroppers, medical monitors, and other sundry devices. I think that the miniaturization trend will continue. But there is a limit to miniaturization. The device must remain large enough to be manipulated. Recently I had a rather sophisticated amplifier constructed, using the most advanced solid state components. Had conventional vacuum tubes been used, the components would have occupied a volume of about 2 cubic feet. The total electronics were about the size of a cigarette package. However, the amplifier had to have a number of switches and dials to control its operation. The smallness of these components is fundamentally limited by the size of the hand. These switches and dials had to be mounted on a panel that is part of the device. In the end this amplifier was not much smaller than an amplifier with conventional electronic components. It was, however, less expensive and lighter.

The videophone is a new device that is being developed for public use in the early 1970s (Figure 13-1). As the name implies, it is a combination of telephone and television. At both the sending and receiving terminals of the system a television camera and receiver are coupled to a conventional telephone. Those conducting the conversation are televised and the visual image is transmitted and displayed simultaneously with the sound. The basic technology for this system is already developed. Most of the effort is now concentrated on making the system more economical and versatile. A number of experimental videophones have already been installed at various locations of the Bell Telephone Laboratories and in selected offices of the Westinghouse Corporation. At Westinghouse 40 sets have been installed in the New York and Pittsburgh offices of the company. At present the system is too expensive for home use. Initially it will be used primarily for business communications. Because of the more intimate contact afforded by the videophone, it is assumed that business travel will be reduced. The savings in time and travel expenses should more than justify the cost of the device.

A number of companies, among them Radio Corporation of America, Columbia Broadcasting System, Sony, and Ampex, are actively involved in developing television tape players for home use. It is estimated that the tape player will cost about $400, and half-hour tapes of various programs will sell for $10. The programs will be viewed on a conventional television set to which the player will be attached. With some of the systems it will also be possible to record programs received by a television antenna.

The television tape players being developed by most of the companies utilize conventional video-taping techniques. The video signal is recorded on magnetic tape from which the signal can be reproduced. The technique used by RCA is novel. Here the visual information is stored holographically on special plastic tape. The holograph varies continuously in accord with the changing images it stores. In the tape player the holograph is reproduced by a small laser. A small television camera photographs the image produced from the holograph and the video signal then drives the television. At this point it is not clear which system is preferable. The conventional systems are of proven reliability, but the tapes are expensive. In the RCA system, the initial recording is relatively expensive, but the replication of tapes is inexpensive.

The RCA system uses holographic image reproduction only as an intermediate step in the production of a conventional television picture. However, experiments are also being conducted in the electronic transmission and reproduction of three-dimensional holographic images. The development of such systems is still in the early stages.

Although we have not discussed computer technology, I think that the capabilities of computers are known to most readers. Computers can store and rapidly process vast amounts of information. As the technology progresses, it will become possible for people with virtually no training to communicate with these machines. Computers coupled to television screens can display data, drawings, and diagrams from which the operator can obtain the desired information. The growth of communication links between computers and the easy human access to this complex information chain will certainly result in more efficient industrial and management procedures.

In this field of computer-man communications, a considerable amount of work has already been done in teaching with computers. I do not think that it is clear at this point whether these teaching techniques are successful. The proponents of the

igure 13-1   Videophone. (Courtesy of American Telephone and Telegraph.)

system point with pride to the speed with which the students assimilate the information taught by the computer. They claim that if the machine will do the job of teaching the routine techniques, such as arithmetic and spelling, the teacher will have more time to guide the student in the more challenging areas of learning. On the other hand, the opponents of the system claim that there is already too much depersonalization in our society, and reducing human contact in yet another area is not desirable. They also point to some experiments which indicate that computer teaching does not fully involve the student in the learning process. The student tends to leave too much to the computer and becomes lazy. Computer teaching will not be accepted by all. I think, however, that we will see an increasing use of computers in this area.

Computers can be programmed to process visual images and bring about a remarkable improvement in the picture quality. We have seen examples of this in the processing of pictures transmitted by satellites. Here pictures, which due to transmission noise were almost undecipherable, were processed by computer to yield respectable images. An example of image enhancement by computer is shown in Figure 13-2. The computer is programmed to remap the picture tones, to enhance boundaries, and to revise the gray scales, which results in the superior picture.

Some computer experts predict that in the near future computers will be available for household use. The home computer will be linked to the others in the bank, library, grocery, or department store. Through these links the computer will be able to retrieve information from libraries, select food, pay bills, and generally relieve us of routine tasks. It is not difficult to sketch a scenario for a future home fully computerized with its television monitor linked to all information sources in the world.

A fascinating possibility in the future is communication with extraterrestial civilizations. Some estimates indicate that in our galaxy alone there may be as many

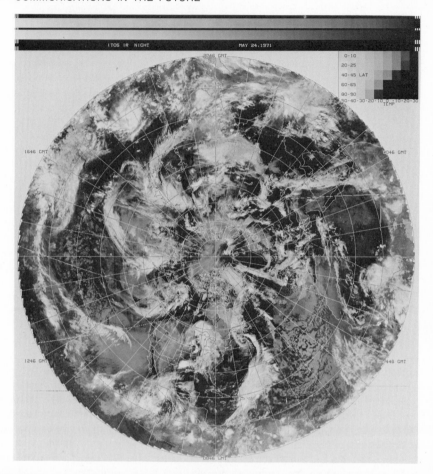

Figure 13-2   An example of picture processing by computer. (Courtesy of NASA.)

as 2 billion stars with planets suitable for intelligent life. The actual number of civilizations with which contact could be made is much more difficult to estimate. Any civilization that has reached a stage where it can start communicating with other civilizations has also reached the stage where it can destroy itself. It is, of course, possible that every civilization destroys itself before it can make contact. Yet with a most optimistic outlook we may suppose that all civilizations work out their initial problems and survive indefinitely. From these two limiting possibilities we conclude that there are between 0 and 2 billion civilizations in our own galaxy able to contact each other.

With our present technology it may be possible to generate signals powerful enough to be marginally detectable at the nearest possible habitated star 10 light years away, but a program for this type of signaling has so far not been started. This is not surprising since the project would be expensive and the prospect for success discouraging. It is not clear in which direction to send the signals and a successful contact could not be confirmed until 20 years from the start of signaling.

Prospects for contact are much better in attempting to detect radio signals that may have been sent by some more advanced intelligence than ours. The development of extraterrestrial radio detection began in 1931 when Karl G. Jansky found radio noise originating from the Milky Way. The source of this signal was natural electromagnetic radiation from the stars. Jansky's observation led to the development of radioastronomy. Following World War II radioastronomy techniques were greatly improved using radar technology and the tracking skills developed for the space program. Radioastronomy is done with a radio telescope, which is a large antenna designed to receive very weak signals. The antenna is usually in the form of a disk 100 feet or more in diameter (Figure 13-3). In order to detect the weak signals highly sophisticated signal-amplifying techniques are used. The radio telescope at Arecibo, Puerto Rico, is capable of detecting artificially generated radio signals from as far away as 100 light years.

There are dozens of radiotelescopes in many parts of the world regularly scanning the skies. The primary task of radiotelescopic observation is not the detection of artificially generated signals; it is rather the general study of stellar radio sources. Still it is possible that during routine surveys a signal from an extraterrestrial intelligence may be detected.

With the modern radiotelescopes, the possibility of detecting extraterrestrial artificial radio signals is no longer in the realm of science fiction. Many scientists have been seriously concerned with the problem and have tried to guess at the type of signals we can expect to receive and how to decode them. Within 100 light years, which is the extent of the listening capacity of present radiotelescopes, there may be as many as 1000 civilizations. Some of these may be transmitting signals in our direction. However, even if we were certain that an extraterrestrial signal is being

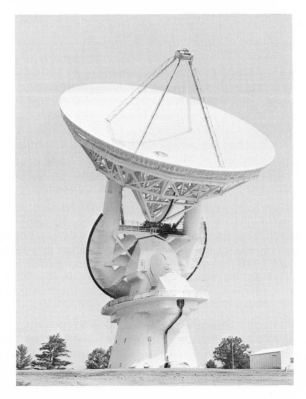

Figure 13-3   Radio telescope. (Courtesy of National Radio Astronomy Observatory.)

sent to us, the problem of detecting it would still be very difficult at this stage perhaps even impossible. It is characteristic of man that at every stage of development he thinks he has reached a near optimum situation. We really cannot be certain that the communication techniques we have developed during the 100 years of our technology are the same as those used by a society possibly millions of years ahead of us in technology. But even if the communication methods are the same, the detection problems are still immense. It is almost certain that communication would be done with a modulated carrier in a narrow frequency range. To detect such a signal, the receiver has to be tuned to the carrier frequency. Scanning the skies for even a single known frequency source is a long and tedious process. A search of all possible transmission frequencies is virtually impossible. A decision therefore has to be made on the most likely frequency used by an extraterrestrial source. The prevalent opinion is that a search at 1420 megacycles would be a reasonable first attempt. This should be a well-known frequency to all intelligences since it is the frequency of the radiation emitted by excited hydrogen which is found in all background cosmic radiation.

A search for artificial signals at this frequency was conducted during May, June, and July in 1960. This search, known as Project OZMA, was carried out with an 85-foot radiotelescope at the National Radio Astronomy Observatory at Greenbank, West Virginia, under the direction of Frank D. Drake. The observations were confined to two stars, Epsilon Eridani and Tau Ceti, both of which are about 11 light years away. The search was conducted during a total of 150 observation hours. I do not think anybody was surprised that the results of the search were negative. Only two stars were under observation at a very small frequency range and only for a very short time. I am certain that more projects of this sort will be conducted in the future.

Beyond the developments we have discussed lie the use of presently unknown techniques. It is possible that entirely new forms and media for communication will be discovered. I have read something along these lines in *Physics Today*, a journal of the American Institute of Physics. A few years ago it had been proposed that there may exist in nature particles which move faster than the speed of light. These hypothetical particles are called *tachyons*, after the Greek word for speed. Although Einstein's theory of relativity shows that particles cannot be accelerated to speeds larger than the speed of light, it seems that particles which always move with speeds larger than that of light are not excluded by theory. There has been much discussion about the feasibility of tachyons, but so far they have not been observed. In connection with these particles Stephen L. Brown of the Stanford Research Institute writes:

> Tachyons could therefore be used for communication systems. Such communication systems would be useful only where ordinary electromagnetic radiation is too slow, as in interstellar communication. Finally, it would seem likely that any extraterrestrial life of high technology would be aware of tachyons (if they exist) and would use them for communications instead of waiting centuries for replies at the speed of light. Perhaps, then, the Project OZMA concept of monitoring electromagnetic radiation for intelligible patterns will turn out to have much less potential for interstellar contact than a tachyon monitoring system.

The exact course of our future technology is not totally predictable. One thing is certain, however, the technology will continue to grow. Whether this technology will improve or degrade the quality of our lives depends on how we use it. If the past is any indicator of the future, it will do both.

DATES OF SIGNIFICANT EVENTS IN COMMUNICATIONS

1600–1750: Initial studies of electric and magnetic phenomena started.
1780–1790: Coulomb experimented on the force between electric charges.
1790–1800: Voltaic cell was discovered.
1820: Oersted showed that electric currents produce magnetic fields.
1830–1840: Faraday and others showed that electric fields are produced by changing magnetic fields.
1834: Gauss and Weber built electromagnetic telegraph.
1837: Cooke and Wheatstone telegraph was built.
1842: Morse's telegraph was installed between Baltimore and Washington.
1860: First telephone was built by Reis.
1864: Maxwell's theory of electromagnetism was published.
1866: Permanent transatlantic telegraph cable was installed.
1876: Telephone was developed by Bell.
1887: Hertz experimented on electromagnetic radiation.
1894: Lodge demonstrated electromagnetic wireless communication over a distance of 150 yards.
1898: Marconi and Jackson transmitted a signal over a distance of 60 miles.
1901: Transatlantic wireless communication was developed by Marconi.
1904: Fleming developed vacuum tube diode.
1906: Lee DeForest invented vacuum tube triode.
1907: Fesseden broadcasted speech over a distance of 200 miles.
1908: Campbell-Swinton published basic ideas on television broadcasting.
1920: First scheduled broadcast was transmitted by station KBKA, Pittsburgh.
1929: Zworykin demonstrated his television system.
1933: Armstrong developed frequency modulation.
1936: Commercial television broadcasting was started by BBC.
1948: Transistor amplifier was built by Brattan, Bardeen, and Shockley.
1950–1960: Microwave communication links were developed.
1954: Color television broadcasting was started in the United States.
1954: The first maser was built.
1960: The first laser was built.
1960: The communication satellite was launched.

Appendix I
Nontechnical Aspects
of Communication

Our modern communication technology has had an enormous impact on the structure and functioning of societies. A study of this aspect of communication, however, is beyond the scope of this book. The references listed below are intended to aid those who want to examine the broader aspects of communications. This is not an exhaustive survey of the literature on these topics and should be regarded only as a starting point for further study.

REFERENCES TO NONTECHNICAL ASPECTS OF COMMUNICATION

Politics

CHESTER, E.W., *Radio, Television and American Politics*. New York: Sheed and Ward, 1969.

MACNEIL, R., *The People Machine*. New York: Harper and Row, 1968.

VERRIER, A., "Hot Line," *New Statesman,* November 15, 1963.

VERRIER, A., "Open Wire to Kremlin," *Business Week,* June 24, 1963.

VERRIER, A., "U.S.-Russian Hot-Line Agreement," *Current History,* September 1963.

LORD WINDLESHAM (D.J.G. HENNESSEY, BARON), *Communication and Political Power*. London: Trinity Press, 1966.

WYCKOFF, G., *The Image Candidates*. New York: Macmillan, 1968.

Violence

BERKOWITZ, L., "The Effects of Observing Violence," *Scientific American,* **210** (1964).

BLUM, S., "De-escalating the Violence on TV," *New York Times Magazine,* December 8, 1968.

CRESSEY, D.R., AND D.A. WARD, *Delinquency, Crime and Social Process*. New York: Harper and Row, 1969.

ELLISON, J., "Stimulant to Violence," *The Nation, 197* (1963).

HALLORAN, J.D., *The Effects of Mass Communication*. Leicester, England: Leicester University Press, 1964.

LARSEN, O.N. (Ed.), *Violence and the Mass Media*. New York: Harper and Row, 1968.

SCHRAMM, W., "The Effects of Television on Children and Adolescents," *UNESCO Courier, 18* (1965).

Eavesdropping

BRENTON, M., *The Privacy Invaders*. New York: Coward-McCann, 1964.

CARROL, J.M., *The Third Listener*. New York: E.P. Dutton, 1969.

LONG, E.V., *The Intruders*. New York: Praeger, 1966.

U.S. House of Representatives, Committee on the Judiciary (1953), *Wiretapping for National Security*.

U.S. Senate, Committee on the Judiciary (1961), *Wiretapping and Eavesdropping Legislation*.

General

BARRETT, M. (Ed.), *Survey of Broadcast Journalism*. New York: Grosset and Dunlap, 1969.

BUZZI, G., *Advertising, Its Cultural and Political Effects*. Minneapolis: University of Minnesota Press, 1968.

GLICK, I., AND S. LEVY, *Living with Television*. Chicago: Aldine, 1962.

KLAPPER, J.T., *Effects of Mass Communication*. New York: Free Press, 1960.

MCLUHAN, M., *Understanding Media: The Extensions of Man*. New York: Signet Books, 1964 (Paperback).

ROSENTHAL, R. (Ed.), *McLuhan: Pro and Con*. New York: Penguin Books, 1968.

SCHRAMM, W., *The Process and Effects of Mass Communication*. Urbana, Ill.: University of Illinois Press, 1955.

SCHRAMM, W. (Ed.), *Mass Communications*. Urbana, Ill.: University of Illinois Press, 1960.

SOPKIN, C., *7 Glorious Days, 7 Fun-Filled Nights*. New York: Simon and Schuster, 1968.

STEINER, G.A., *The People Look at Television*. New York: Knopf, 1963.

Appendix II
An Evaluation of Extrasensory Perception (ESP)

Having discussed conventional methods of communication, we will now turn to the rather controversial subject of extrasensory perception (ESP). Many of the proponents of ESP have suggested that ESP could be used for communication. Indeed if ESP were a real phenomenon it would be a superb means of communication. Some of the reported occurrences indicate that the transmission of ESP messages is not affected by distance or any material between the sender and receiver. This alleged property of ESP would be very useful for submarine communication because water severely impedes the propagation of conventional signals. It is therefore not surprising that the U.S. Office of Naval Research supported some of the ESP experiments conducted at Duke University by J.B. Rhine.

In this chapter we shall attempt to assess some of the evidence for ESP.

Extrasensory perception has been categorized into four processes:

1. Telepathy: a person's awareness of another person's thoughts without communication through any human sensory channels.
2. Clairvoyance: knowledge acquired of an object or an event without the use of the human senses.
3. Precognition: a person's knowledge of another person's future thoughts or events.
4. Psychokinesis: a person's ability to influence a physical object or an event by his thoughts.

The interest in ESP is probably as old as man. In many ways it is related to the quest of man for religion or something in himself that transcends physical laws. In his book, *Parapsychology*, J.B. Rhine, a foremost experimenter in this field in the United States, writes,

If it is correct to define parapsychology as the science dealing with nonphysical personal agency, it is hard to see what legitimate problem or claim of religion would not, if it were brought to the point of careful investigation, belong to the domain of that science. This would make the relation of parapsychology to religion something like that of physics to engineering or biology to medicine.

The evidence offered for ESP falls into two categories: (1) strange events called *spontaneous events* and (2) actual laboratory ESP experiments. The spontaneous events consist of reports about abilities of certain people to foretell events, to communicate by telepathy, or to communicate with the dead. Laboratory experiments have been done on all four aspects of ESP. The experiments are all statistical in nature. Most of the telepathy, clairvoyance, and precognition experiments are done with cards. A deck usually containing 25 cards with five different symbols is shuffled, and the subject then attempts to guess the identity of a given card. Without any ESP a subject would, on the average, guess correctly 5 out of the 25 cards. An average guess of better than five indicates that something other than chance is in action, possibly ESP. Experiments in psychokinesis are usually done with dice. Here the subject tries to influence the fall of the die on a particular face.

We will try to evaluate both the spontaneous and laboratory ESP evidence. But before we do this, I must state my own attitude toward ESP. This is important because the conclusions people draw from ESP experiments and reports depend strongly on their initial attitude toward the subject. Even the results of experiments seem to depend on whether the experimenter believes in ESP or not.

I do not think anyone approaches ESP without a bias. The attitude toward it is too closely bound up with the general outlook we all carry with us. Basically, I did not believe in ESP, and having done much reading on the subject, I still do not believe in it. There has been a great deal written on ESP and it has not been possible for me to survey the whole literature. I think, however, that I have come in contact with most of what is significant. I believe that at present there is no evidence for the existence of ESP. But to be fair, I must point out that there are very capable people who have read the same articles and have come to different conclusions.

SPONTANEOUS EVENTS

We will first talk about spontaneous events. In this category we will first examine some of the strange tales that frequently crop up. These are wonderful to listen to but are almost impossible to check. They are usually told by a single individual, so it is not really possible to decide whether he is fabricating a story, has had a hallucination, or whether the tale was a real occurrence. In cases where reliable checks were possible these stories turned out to be fictitious or distorted to such an extent that when the distortions were removed, the supernatural aspect of the story disappeared.

One of the most impressive tales, because of the respectability of the teller, was described by Gurney and Myers in the magazine *Nineteenth Century,* **XVI** (July 1884). The event was reported to have happened to Sir Edmund Hornby who was the Chief Judge of the Supreme Consular Court of China and Japan at Shanghai. The story as told by Hornby and the subsequent analysis is in Hansel's book (see References at the end of this appendix). It is one of the few stories of this type that comes from a reputable source and has been carefully checked. The following is an abbreviated version of the story which in the original was told in the first person by the judge.

Usually in the evening Judge Hornby wrote his judgments at home and gave them to his butler, who in turn was to give them to the news reporters. On the night of the occurrence Sir Hornby gave his judgments to the butler and went to sleep before 12 o'clock. He was awakened by a tapping on his door. The visitor was a reporter, who in the original recounting is called Mr. X. The judge was very annoyed by the intrusion, but there was something in the reporter's manner and appearance that made the judge decide not to eject him by force. The reporter looked deathly pale and as if in pain. Hornby glanced at the clock, noted that it was about 1:20 A.M., and said, "The butler has had the judgment since half-past eleven; go and get it." The reporter replied, "Pray forgive me; if you knew all the circumstances, you would. Time presses. Pray give me a precis of your judgment, and I will make a note in my book of it." After some arguing the judge decided to give him a summary of his judgment because he did not want the conversation to waken and frighten his wife. The reporter thanked the judge and left. The time was exactly 1:30 A.M.

The following morning Sir Hornby told the story to his wife who said that she did wake up and thought she heard talking. The judge asked the butler if he had let anyone in the house the previous night. The butler said that he did not. The doors were locked from the inside in the evening and were still locked that way in the morning.

When the judge went to court that morning he was told by the usher that reporter X was found dead in his room. The reporter's wife had left him in his room at about midnight. When she returned at 1:30, he was dead. She called a doctor who arrived at 2 o'clock and concluded that the reporter had been dead for about one hour. In the reporter's notebook there was writing which began, "The Chief Judge gave judgment this morning in the case to the following effect . . ." What followed was in undecipherable shorthand. According to the judge, he ordered an inquest. It was found that the reporter had died of a heart disease, and he could not have left the house without being observed between 11 and 1 o'clock on the night of his death.

Sir Hornby did not reveal his story to the public for nine years. He later wrote:

As I said then, so I say now — I was not asleep, but wide awake. After a lapse of nine years my memory is quite clear on the subject. I have not the least doubt I saw the man — have not the least doubt that the conversation took place between us.

Next November in the same magazine a letter by F.H. Balfour pointed out some discrepancies in this story: The reporter was the Reverend Hugh Lang Nivens, editor of the *Shanghai Courier*. He died not at one o'clock in the morning but between 8 or 9 A.M. At the time of the event there was no Lady Hornby. Sir Edmund's second wife had died two years earlier and he did not marry again till three months after the event. Contrary to the judge's story, no inquest was ever held. The story centers on a judgment that was to be delivered the following day, January 20, 1875. There was no record of such a judgment.

Before publishing the letter Judge Hornby was asked to comment on it. He wrote:

My vision must have followed the death (some three months) instead of synchronizing with it. At the same time this hypothesis is quite contrary to the recollection of the facts both in my own mind and in Lady Hornby's mind. . . . If I had not believed, as I still believe, that every word of it [the story] was accurate, and that my memory was to be relied on, I should not have ever told it as a personal experience.

There are numerous other stories of the supernatural that have been claimed to occur. They all have this in common: When they can be checked, it is found that many facts have been distorted, time sequences have been rearranged, and when the episode is reconstructed, the supernatural flavor disappears.

In the past much more attention was paid to these types of tales since people were willing to devote the time for an honest and thorough investigation of such reports. Today reports of such occurrences as evidence for the existence of supernatural events are usually dismissed.

Most people have had experiences which, if they were so inclined, they could attribute to ESP. Premonitions and simultaneous thoughts with another person have been experienced by many people. But these types of events cannot be considered as evidence for ESP. They are more logically attributed to coincidence. We seldom remember the premonitions that do not materialize, and it is easily conceivable that a stimulus triggers the same thought sequence in two people.

Spiritualism is another frequently encountered ESP event. As the name implies, spiritualism is communication with the dead. A medium claims to have a spirit contact in the other world. This contact, on being aroused by the medium, can perform supernatural acts. The type of performance that was popular before World War II was the seance which usually took place in near darkness. One or two people, called *controls*, usually sat next to the medium to make certain that the medium was not responsible for tricks. After the medium established contact with the spirit, all sorts of peculiar things happened. Voices were heard, tables levitated, cabinets were turned over, and so on.

One of the most famous mediums of this type was Margery Crandon who was active in the 1920s and 1930s. Margery Crandon was the wife of a well-known Boston surgeon who was present at most of her seances and was thought by many to be responsible for most of the occurrences during the seance. As was usual her seances were given in near darkness. When she went into her trance, she claimed to be in touch with Walter, her deceased brother, who then caused the extraordinary events.

Margery Crandon's performance was taken rather seriously and was investigated by many groups. The first investigation was done by a group of four Harvard psychologists, McDougall, Roback, Murphy, and Helson. They were not impressed favorably and did not issue a report on their investigation. At about this time *Scientific American* offered $5000 to anyone who could exhibit genuine psychic phenomena. An investigating committee was formed. The committee was convinced that Margery's powers were genuine. Margery was about to receive the prize when Harry Houdini was invited to a seance and detected trickery. He was able to show that in the near complete darkness of the seance Mrs. Crandon was able to move objects with her feet and her head while her hands were held by the control agent.

Despite the discovery of trickery on her part, groups continued to study her. She was investigated by E.J. Dingwall of the British Society for Psychical Research. Although he believed that most of the things Margery did could have been produced by trickery, he did not want to make a definite judgment on her. Next, Hoagland, a psychologist at Harvard, and four of his associates studied her. In 1925 they published an article in the *Atlantic Monthly* showing that she was a trickster. Extensive studies have been made also on other mediums with similar results.

Margery was a physical medium. Her performance consisted mostly of objects being mysteriously moved. Today most of spiritualism centers about mental mediums. In many cases actual contact is claimed by the medium with some deceased person. Through this contact, communication is established with other spirits of interest to the person who comes to consult the medium. The proof for the genuineness of the contact is the fact that information is revealed which could only be known to the dead person and the client. This type of spiritualism exists today and probably always will. In the late 1960s there were reports in the papers about Bishop Pike contacting his son who committed suicide. Bishop Pike subsequently described his experiences in a moving article published in two issues of *Look* magazine. Following his son's death a number of strange events occurred which suggested to him that his son continued to interact with him. Pike visited a number of mediums through whom he believed he established communication with a number of deceased people, including his son.

These reports are nearly impossible to check reliably, however. The reports are usually from distraught people who come for consolation. They want very much to believe that the loved person is not completely gone. It is not possible to know how much information the medium has obtained prior to the seance or to what extent the anxious person has subconsciously helped the medium by his reactions as the medium probes toward the right answers. Skeptics who have gone to see these mediums report that their answers tend to be vague, probing, and so constructed that it is possible to read anything into them.

There have been many supernatural phenomena reported from other cultures. Most familiar of these are the reports of Yoga events from India. There are many tales of persons levitating, that is, raising themselves up from the ground in defiance of gravity. If such a thing could be reliably demonstrated, it would be the most revolutionary discovery in physics. The writer Arthur Koestler went to India and Japan in 1959 specifically to find out about some of the mysteries of the East. One of the things he tried to investigate was levitation. In India there are a number of institutes that study the ability of people who are well versed in Yoga. He visited these institutes. Some of the researchers there firmly believed there are people who levitate. Koestler, however, found no reliable confirmation of this. One would think that levitation would be easy to demonstrate. A person who is supposed to have the ability is placed on a scale. It is not even necessary that he levitate completely. The effect would be demonstrated if he could make himself just a little lighter. None of the institutions have reported such an experiment.

EXPERIMENTS ON ESP

We will now examine some of the experiments on ESP. There are many. They are usually of the card-guessing type and most of them show no evidence for ESP. But some do show successful guesses that are clearly higher than would be expected statistically from chance guessing. It is clear that in these experiments something other than chance guessing was taking place. There are only two explanations for the results of these experiments: either the subject has ESP abilities or there has been conscious or subconscious cheating during the experiments.

It may seem entirely unfair to apply the criterion of cheating to ESP experiments. If we applied this criterion to other fields of experimental work, we could dismiss the results of any experiment by supposing that it was a fraud. But there are some important features of ESP experiments that separate them from other fields.

They are not consistently reproducible. In the middle 1930s after Rhine reported his experimental results showing evidence for ESP, there were numerous attempts to reproduce his results. Some indeed reproduced them, but many did not. In 1936 W.S. Cox of the psychology department at Princeton University tested 132 subjects producing 25,064 trials with no evidence for ESP. E.T. Adams of Colgate University reported results of 30,000 trials; J.E. Crumbaugh from Southern Methodist University tested 100 subjects and recorded 75,600 trials; Raymond W. Houghby of Brown University recorded 41,250 trials; and C.P. and J.H. Heinlein of Johns Hopkins recorded 127,500 trials; none of these showed evidence for ESP and there were many others.

When I mentioned the possibility of trickery in these experiments, I did not mean to imply necessarily that the people who conducted these experiments cheated. It is possible that the subjects of the experiments may have used conscious or subconscious tricks that fooled the experimenters. Rhine himself mentioned that when controls and precautions are tightened, the incidence of ESP goes down. He writes, "Elaborate precautions take their toll. Experimenters who have worked long in this field have observed that the scoring rate is hampered as the experiment is made complicated and slow moving. Precautionary measures are usually distracting in themselves." Rhine interprets the fact that precautionary measures reduce the incidence of ESP by saying that they put the subject in the wrong state of mind. Another interpretation is that precautions simply reduce the possibility of cheating.

Rhine also found that there is a better chance for obtaining ESP results if both the subject and the experimenter believe in ESP. He attributes this to a more favorable mental state which exists under these conditions. Another explanation may be that this type of person is more likely to use tricks consciously or subconsciously to bias the experiments in favor of ESP since he has a point to prove. In ESP there is a complete inability to predict future results from past experiments, a very important requirement for any experimental work. Subjects who have shown high ESP abilities for one experimenter show average guessing for another.

It is not possible for us to discuss all of the experiments that claim to show the existence of ESP. Many of these experiments have been rejected by Rhine and others as not having been done under sufficient controls. Hansel in his book has taken a number of experiments that are considered by workers in the field as the most conclusive. He has analyzed them in detail and shown that in each case trickery could have been employed. It can never be shown conclusively that trickery was used because these experiments were done in the past, but his analyses do offer an alternate explanation to ESP. The analyses are rather dull. They involve a detailed explanation of all the procedures in the experiments and then a demonstration of where cheating could have occurred. In some cases Hansel has repeated the experiments and showed that his hypothesis is likely. We will discuss briefly one of the many experiments analyzed by Hansel. This is the series of experiments by Pearce and Pratt. Only the salient features of Hansel's presentation will be given.

The Pearce and Pratt experiments were done in 1933 and 1934 under some supervision by Rhine. These were long-distance clairvoyance experiments in which Pearce, who was a divinity student, guessed at cards from a pack controlled by Pratt, who was a graduate student in psychology. The two of them were in different buildings of the Duke University campus. Their watches were synchronized. At the fixed time Pratt took the first card from a shuffled ESP deck which was on his right side and placed it face down on a book at the center of the table. After one minute he transferred this card still face down to the left side of the table and placed another card on the book. In this way Pratt went through two card decks (50 cards). At the end of the sitting he recorded in order the identity of the cards, sealed the list in an envelope, and delivered it to Rhine.

In the other building Pearce recorded his sequential guesses. He also placed his list in an envelope and delivered it to Rhine for comparison. The results of these experiments were truly remarkable. For example, in one set of 300 trials Pearce obtained 119 hits. This is equivalent to an average guess of 9.9 hits out of 25 trials.

Remember than an average guess of 5 out of 25 would be the expected result from random guessing. The probability of obtaining Pearce's results by chance is the incredibly small 10^{22} to 1. This is so small that we cannot attribute Pearce's success to chance. We are faced with two alternatives, either Pearce had ESP abilities or there was cheating during the experiments. Rhine and Pratt discussed these experiments in a 1954 issue of the *Journal of Parapsychology* in which they state these two alternatives and point out that if cheating took place, then all three of them, Rhine, Pratt, and Pearce, had to cheat. The line of attack that Hansel takes is to show that not all three had to cheat. To explain the results it is sufficient to assume that Pearce cheated. Hansel then shows that Pearce did indeed have the opportunity to cheat. In 1960 Hansel went to Duke University and examined the location and procedure of the experiments. Recall that after each run Pratt turned over the cards from the deck and recorded their identity. The room in which Pratt sat had a number of clear glass windows as well as clear glass windows above the doors. Hansel postulates that if Pearce had chosen to cheat, he could have returned unobserved to the building where Pratt sat and looked through one of the openings as the cards were examined by Pratt. A number of details which Hansel gives make this hypothesis plausible. Hansel himself conducted an experiment while there. He asked W. Saleh, a member of the research staff at Duke to go through a deck of ESP cards as was done in the experiment. Saleh was in the room originally used by Pratt. Without Saleh's knowledge Hansel stood on a chair and looked into the room through a crack in the door. As Saleh was recording the cards, Hansel was able to see them and obtained 22 hits out of 25.

We will probably never know with absolute certainty whether cheating did occur. However, Hansel's assumption offers a very strong alternative to ESP.

There have been a number of experiments that claim to show evidence for psychokinesis (PK). Martin Gardener, in his book *Fads and Fallacies,* describes a PK experiment done by Richard S. Kaufman in 1952 at Yale University. This was a dice-tossing experiment in which eight people participated. Four of them believed in PK, the other four did not. Each person kept records of 40 tosses. The believers recorded results which were in favor of PK. The recordings of the nonbelievers showed the opposite. Unknown to the tossers, the experiment was recorded by a hidden camera. The film showed that the believers made errors favoring PK and the nonbelievers made opposite errors. The actual results were entirely in accord with chance. These wrong readings were not necessarily due to fraud. In the excitement of the experiments mistakes were made and they were guided by the bias of the experimenters.

Many critics have suggested experiments that would convince anyone of the reality of ESP if it indeed exists. Hansel suggested that since Rhine claims that one person in five displays ESP ability, it should be relatively simple to select 100 subjects with good ESP potential. The skeptic would then think of one of five symbols and the 100 subjects would each press one of five push buttons corresponding to the symbols used in the experiment. If the subjects do possess ESP abilities, then the symbol with the largest number of votes should correspond to the symbol picked by the skeptic in most trials.

To my knowledge there has been only one experiment of this type reported. It was described by Rhine (see References):

> One of the other types of approach to the control problem is the effort to use statistical methods of concentrating small bits of information into more recognizable magnitude. Early in the Duke work the effort was made to use repeated guessing of cards individually enclosed in sealed opaque envelopes to see whether a compounding of responses of a given type would occur and, if so, whether it would tell with relative certainty the true nature of the enclosed target. As already stated, the psi-missing factor plays havoc with this thoroughly logical approach. This factor, along with the fact that the operation of psi is unconscious and, in addition, that the subject's performance varies greatly as he goes through the test day after day, introduces irregularities that disqualify this repeated-guessing method taken alone.

The mysterious psi-missing factor to which Rhine attributes the failure of this experiment is a general escape clause which Rhine and others use when their ESP experiments give negative results.

Critics of ESP have repeatedly called for experiments that would mechanize the selection of targets and the recording of guesses. Such an experiment was conducted at the U.S. Air Force Research Laboratories. A machine randomly generated numbers from 0 to 9 and the guesses of subjects were recorded by the apparatus. Experiments were done on clairvoyance, precognition, and telepathy. Thirty-seven subjects were tested with a total of 55,500 trials. No evidence was found for ESP.

CONCLUSIONS

The picture I have given here is necessarily one-sided; as I have mentioned before there are many people who have thoroughly examined the subject and have come to the conclusion that there is indeed something to ESP. Arthur Koestler whose works I respect wrote an article on ESP in the May 7, 1961, issue of the *Observer* (London). He believes that the existence of ESP has been conclusively demonstrated. He is convinced that the ESP experimental techniques are rigorous and scientific. Koestler is very critical of scientists who refute ESP. He compares them to the orthodox Aristotelians of the middle ages, who refused to accept new ideas because they represented a threat to their establishment. He quotes H. J. Eysenck, a professor of psychology at the University of London:

> The very possibility of extrasensory perception or psychokinesis appears contrary to modern scientific logic and many people have shown considerable reluctance even to look at the evidence that has been produced in favor of these alleged abilities. Scientists especially when they leave the particular field in which they have specialized are just as ordinary pig-headed and unreasonable as anybody else and their unusually high intelligence only makes their prejudices all the more dangerous.

Koestler's article is very general. He gives no specific examples to support his point of view. In that issue of the *Observer* there was a comment that letters to the editor concerning Koestler's article would be published the following week and Koestler's reply the week after.

The following week many letters were published both for and against Koestler's article. There were many important criticisms of Koestler's thesis including a letter from Hansel. Koestler's statements were challenged, specifically his claim that ESP experiments were rigorous and scientific. It was pointed out that in 1939 the American Psychological Association stated the conditions that should be observed in ESP experiments. No experiments were done that way. Koestler was asked to produce one conclusive experiment. The week after there was a letter from Koestler. However, he did not answer any of the criticisms that challenged his view of ESP.

I wrote a letter to Koestler in the summer of 1968 and asked him if now, eight years later, he still believes in ESP. A few weeks later Koestler wrote me a brief letter in which he stated without hesitation that he thinks that ESP is a real phenomenon. He referred my letter to Mrs. Rosalind Heywood who is a council member of the Society for Psychical Research in England. I have since received a number of letters from her in which she explained to me her views on ESP. She also sent me a number of interesting references that pointed out some of the recent trends in ESP. These trends, I think, are shown in a book called *Science and ESP*. This book contains articles by 13 authors with somewhat different attitudes to ESP. Some believe in all aspects of ESP, others express some skepticism. The articles are in the nature of philosophical speculations. There is virtually no discussion of ESP experiments. Some of the people who are seriously involved with ESP now question the validity of the very specific card-guessing experiments. They claim that a phenomenon as ethereal as ESP cannot be expected to exhibit itself in guessing mundane cards. Possibly the place to look for ESP is in the dream state. ESP should not be expected to duplicate information but rather it should produce some sympathetic emotions and images. Many experiments of this type are in progress.

References

BROAD, C. D., *Lectures on Psychical Research*. London: Routlige & Kegan Pave, 1962.

HANSEL, C. E. H., *ESP, A Scientific Evaluation*. New York: Scribner, 1966. This book contains a large list of other references. Reviews and criticisms of Hansel's book are found in the *Journal of the American Society for Psychical Research*, July 1967, and *Journal of the Society for Psychical Research*, March 1968.

RHINE, J. B. (Ed.), *Parapsychology, from Duke to FRNM*. Durham, N.C.: Parapsychology Press, 1965.

RHINE, J. B., and J. G. PRATT, *Parapsychology*. Springfield, Ill.: Thomas, 1957.

SCHWARZ, B. E., *A Psychiatrist Looks at ESP*. New York: Signet Mystic Book, 1965.

SMYTHIES, J. R. (Ed.), *Science and ESP*. New York: Humanities Press, 1967. This book contains a large list of references.

Appendix III
Review Problems

2-1 Summarize the knowledge of electricity as of about 1750.

2-2 Explain the observations about electricity in terms of the simple atomic theory discussed in the text.

2-3 What is Coulomb's law?

2-4 What are "lines of force?"

2-5 By connecting a very large conducting plate to the earth, it is possible to shield out the effect of electric charges (see the drawing).

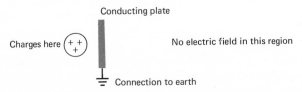

(a) Explain how shielding occurs. (*Hint:* The earth is a very good source of electric charge.)

(b) Why is it not possible to shield out gravitational forces?

2-6 Sketch the electric field produced by two positive charges fixed to be at a distance *d*.

2-7 Explain the fact that a balloon which has been rubbed by a dry object will stick to a wall but will not stick to a metal surface.

2-8 Using our simple theory of the atom, which of the elements lithium, beryllium, boron, carbon, nitrogen, oxygen, and flourine would you expect to be chemically most reactive? (Refer to the periodic table of the elements.)

2-9 Lithium flouride (LiFl) is a very stable salt, much like the common table salt (NaCl). Why is the molecule of this salt so stable?

2-10 State Oersted's observation.

2-11 State Faraday's observation.

2-12 Describe briefly how the following devices work:
   (a) Voltaic cell
   (b) Ammeter
   (c) Generator
   (d) Transformer
2-13 Explain why some materials are magnetic.
2-14 Consider the following circuit:

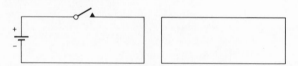

   (a) Sketch the magnetic field around the wire of loop 1.
   (b) The switch in loop 1 is opened, which interrupts the flow of current.
        Using Lenz's law deduce the direction of the induced current in loop 2.
        (Indicate direction of conventional current.)

CHAPTER 3

3-1 Describe briefly the operation of the telegraph system designed by (a) Cooke
     and Wheatstone; (b) Morse.
3-2 State briefly the principles on which the components of the systems of Problem
     3-1 are based.

CHAPTER 4

4-1 Explain briefly the nature of sound.
4-2 Explain briefly the operation of (a) a microphone, (b) a speaker.
4-3 Describe the operation of a simple telephone system.

CHAPTER 5

5-1 Explain briefly how an electromagnetic wave is generated.
5-2 Illustrate the concepts of frequency, wavelength, and period of a wave. What
     are the relationships between these?
5-3 Sketch a simple radio system of the type used by Marconi and explain its
     operation.

CHAPTER 6

6-1 Explain briefly the following: force, energy, work, and power.
6-2 Define current, voltage, and electrical resistance, and state the phenomeno-
     logical law that relates these quantities (Ohm's law).
6-3 Describe the function of a capacitor, an inductor, and a resistor in electric
     circuits.
6-4 Why does an electric light usually burn out when it is first switched on?
6-5 Derive the expression for electric power in terms of voltage and current,
     current and resistance, resistance and voltage.
6-6 Explain what is meant by direct current and alternating current.

6-7   Explain the operation of a resonant circuit.

6-8

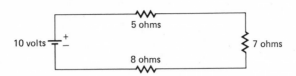

Calculate
    (a) The current flowing in the above circuit.
    (b) The voltage across each resistor.
    (c) The power dissipated by each resistor.
    (d) The total power delivered by the battery.

CHAPTER 7

7-1   Discuss briefly the problem of detecting radio waves.

7-2   Explain briefly the operation of the following devices: (a) diode detector, (b) triode amplifier, (c) triode oscillator.

7-3   Sketch a block diagram of an amplitude-modulated broadcasting and receiving system. Describe the function of each component block.

7-4   Describe the principles of frequency-modulated broadcasting and receiving.

CHAPTER 8

8-1   Describe the operation of a tape recorder and a phonograph.

8-2   Describe briefly what is meant by the "frequency response" of a system, and explain why the knowledge of it is important in the evaluation of a sound-reproducing system.

8-3   Explain briefly what is meant by the "dynamic range" of a sound-reproducing system.

8-4   Explain the principle of stereophonic sound reproduction.

CHAPTER 9

9-1   Explain briefly the operation of a facsimile transmission system.

9-2   Describe the operation of an iconoscope scanner.

9-3   Describe the operation of a cathode-ray tube (picture tube).

9-4   Sketch the block diagram of a television system and explain the function of each block.

9-5   Explain briefly the operation of color television.

CHAPTER 10

10-1 Describe briefly the mechanism of current conduction in (a) metal, (b) pure semiconductor, (c) n-type semiconductor, (d) p-type semiconductor.

10-2 Explain briefly the operation of a semiconductor diode.

10-3 Explain briefly the operation of a transistor.

CHAPTER 11

11-1   Explain the advantages of communications with high-frequency carriers.

11-2   Explain briefly the principles of multiplexing.

11-3   Describe the operation of a klystron.

11-4   Describe the operation of radar.

CHAPTER 12

12-1   Describe the properties of light produced by (a) an incandescent lamp, (b) a spectral light source, and (c) a laser.

12-2   Describe the interaction of light with atoms and explain (a) absorption, (b) spontaneous emission, and (c) stimulated emission.

12-3   Explain the operation of a laser.

12-4   Explain briefly the principles of holography.

References

The numbers in brackets indicate the chapters to which the references are relevant.

ASHFORD, T. A., *The Physical Sciences*. New York: Holt, Rinehart and Winston, 1967. [Chapter 2]

BECK, A. H. W., *Words and Waves* (World University Library). New York: McGraw-Hill, 1967 (Paperback). [Chapters 1, 3, 4, 5, 7, 8, 9, 11]

BENNETT, W. R., and J. R. DAVEY, *Data Transmission*. New York: McGraw-Hill, 1965. [Chapters 1, 2, 3]

BITTER, F., *Magnets* (Anchor Science Series). New York: Doubleday, 1959 [Chapter 2]

BLOSS, R. W., *Pony Express—The Great Gamble*. Berkeley: Howell-North, 1959. [Chapter 1]

BROWN, E. B., *Modern Optics*. New York: Reinhold, 1965. [Chapters 9, 12]

CHIPMAN, R. A., "DeForest," *Scientific American,* March 1965. [Chapter 7]

DAVID, E. E., "The Reproduction of Sound," *Scientific American,* August 1961. [Chapter 8]

DAVIS, H. M., "Radio Waves and Matter," *Scientific American,* September 1948. [Chapter 11]

DE FOREST, L., *Father of Radio*. Chicago: Wilcox and Follett, 1950. [Chapter 7]

DE SANTILLONA, G., "Volta," *Scientific American,* January 1965. [Chapter 2]

FABRE, M., *A History of Communication*. New York: Hawthorn Books, 1963. [Chapter 1]

FINK, D. L., and D. M. LUTYENS, *The Physics of Television* (Anchor Science Series). New York: Doubleday, 1960 (Paperback). [Chapter 9]

FORD, K. W., *Basic Physics*. Waltham, Mass.: Blaisdell, 1968. [Chapters 2, 5, 6, 7, 12]

GAMOW, G., *Thirty Years That Shook Physics: The Story of Quantum Theory* (Anchor Science Series). New York: Doubleday, 1966 (Paperback). [Chapter 12]

GRIFFIN, D. R., *Echoes of Bats and Men* (Anchor Science Series). New York: Doubleday, 1959 (Paperback). [Chapter 11]

HARMON, L. D., and K. C. KNOWLTON, "Picture Processing by Computer," *Science,* April 4, 1969. [Chapter 13]

HERRIOTT, D. R., "Applications of Laser Light," *Scientific American,* September 1968. [Chapter 12]

HUANG, SHAN, "Life Outside the Solar System," *Scientific American,* April 1960. [Chapter 13]

KIRBY, R. S., S. WITHINGTON, A. B. DARLING, and F. G. KILGOUR, *Engineering in History.* New York: McGraw-Hill, 1956. [Chapters 1, 2, 3]

KOCK, W. E., *Sound Waves and Light Waves* (Anchor Science Series). New York: Doubleday, 1965 (Paperback). [Chapter 4, 8]

KONDO, M., "Faraday," *Scientific American,* October 1953. [Chapter 2]

LEITH, E. N., "Photography by Laser," *Scientific American,* June 1965. [Chapter 12]

LESSING, L. P., "Armstrong," *Scientific American,* April 1954. [Chapter 7]

MAC GOWAN, R. A., and F. I. ORDWAY, *Intelligence in the Universe.* Englewood Cliffs, N.J.: Prentice-Hall, 1966. [Chapter 13]

MILLER, E., "Communication by Lasers," *Scientific American,* January 1966. [Chapter 11, 12]

MORRISTON, P., and E. MORRISTON, "Hertz," *Scientific American,* December 1957. [Chapter 5]

NELSON, D. F., "Modulation of Laser Light," *Scientific American,* June 1968. [Chapter 12]

NEWMAN, J. R., "Maxwell," *Scientific American,* June 1955. [Chapter 5]

OVENDEN, M. W., *Life in the Universe* (Anchor Science Series). New York: Doubleday, 1961 (Paperback). [Chapter 13]

PAGE, R. M., *The Origin of Radar* (Anchor Science Series). New York: Doubleday, 1962 (Paperback). [Chapter 11]

PENNINGTON, K. S., "Advances in Holography," *Scientific American,* February 1968. [Chapter 12]

PIERCE, J. R., "Microwaves," *Scientific American,* August 1952. [Chapter 11]

PIERCE, J. R., *Quantum Electronics* (Anchor Science Series). New York: Doubleday, 1966 (Paperback). [Chapters 10, 12]

PIERCE, J. R., *Waves and Messages* (Anchor Science Series). New York: Doubleday, 1967 (Paperback). [Chapters 7, 11]

ROCKETT, J. M., "The Transistor," *Scientific American,* September 1948. [Chapter 10]

SCOTT, J. B., "Lasers and Holography: New Approach for a Television Tape Play," *Microwaves,* December 1969. [Chapter 13]

SHAMOS, M. H. (Ed.), *Great Experiments in Physics.* New York: Holt, Rinehart and Winston, 1959. [Chapters 2, 5, 7]

SINGER, C., E. G. HOLMYARD, A. R. HALL, and T. I. WILLIAMS (Eds.), *A History of Technology* (Vols. 4 and 5). New York: Clarendon Press, 1958. [Chapters 2, 3]

SLURZBERG, M., and W. OSTERHELD, *Essentials of Radio.* New York: McGraw-Hill, 1948. [Chapters 4, 6, 7]

STILL, A., *Communication through the Ages.* New York: Murray Hill Books, 1946. [Chapters 1, 2, 3, 4, 5, 7, 9]

WARTERS, W. D., "Communication in the Future," *IEEE Student Journal,* March 1969. [Chapter 13]

WHITE, H. E., *Modern College Physics.* New York: D. Van Nostrand Co., 1956. [Chapters 4, 8]

WILLSON, M., "Joseph Henry," *Scientific American,* July 1954. [Chapter 2]

(NO AUTHOR) *The Way Things Work.* New York: Simon and Schuster, 1967. [Chapter 8]

Index